PENGUIN BOOKS

Latitude

'A rollicking story of adventure and scientific exploration.
As gripping as any novel. Full of big ideas and bigger
personalities. Sparkles with intelligence and wit'
Alex Preston, author of *Winchelsea*

'Crane has a rare knack for showing people things they really
ought to see across space and time without them having to get
out of their chair. *Latitude* applies his trademark blend of pace,
rigour and attention to enticing detail, in order to fill in a key
segment of historical and geographical knowledge'
Joe Smith, director of The Royal Geographical society

'A thrilling story of courage, survival and science.
It's an extraordinary, visceral and vivid read'
Geographical Magazine

'A story for our times'
Eastern Daily Press

'Written with Crane's customary blend of
expertise and elan, *Latitude* plots the roots
of modern geography and expeditionary science'
Professor Robert Mayhew

'Terrific'
The Spectator

'An amazing story'

ABOUT THE AUTHOR

Nicholas Crane was born in seaside Hastings, grew up in rural Norfolk and learned winter mountaineering in snowy Scotland. Between 2015 and 2018, he was president of the Royal Geographical Society. He is an award-winning writer and geographer who is well known for his television work as lead presenter for the BAFTA-winning series *Coast*, *Great British Journeys*, *Map Man*, *Britannia* and *Town*. He is the author of more than ten books that include: *Clear Waters Rising*, *Two Degrees West*, *Mercator: The Man Who Mapped the Planet*, *The Making of the British Landscape* and *You Are Here, A Brief Guide to the World*. He has written for the *Daily Telegraph*, the *Guardian* and the *Sunday Times*. Nick has travelled in all seven of the world's continents. With his cousin, Dr Richard Crane, he identified and visited for the first time the geographical Pole of Inaccessibility, the point on the globe most distant from the open sea. He lives in London.

Latitude

*The Astonishing Adventure that
Shaped the World*

NICHOLAS CRANE

PENGUIN BOOKS

PENGUIN BOOKS

UK | USA | Canada | Ireland | Australia
India | New Zealand | South Africa

Penguin Books is part of the Penguin Random House group of companies
whose addresses can be found at global.penguinrandomhouse.com.

First published by Penguin Michael Joseph 2021
Published in Penguin Books 2022
001

Copyright © Nicholas Crane, 2021

The moral right of the author has been asserted

Epigraph from *Candide, or Optimism* by Voltaire. Translation Copyright by Theo Cuffe 2005,
published by Penguin Classics 2005.
Reproduced by permission of Penguin Books Ltd. ©

Printed and bound in Great Britain by Clays Ltd, Elcograf S.p.A.

The authorized representative in the EEA is Penguin Random House Ireland,
Morrison Chambers, 32 Nassau Street, Dublin D02 YH68

A CIP catalogue record for this book is available from the British Library

ISBN: 978–1–405–94734–3

www.greenpenguin.co.uk

MIX
Paper from
responsible sources
FSC® C018179
www.fsc.org

Penguin Random House is committed to a
sustainable future for our business, our readers
and our planet. This book is made from Forest
Stewardship Council® certified paper.

'Only after twenty-four hours did they see daylight again; but their canoe smashed to pieces in the rapids; they dragged themselves from boulder to boulder for an entire league; finally they emerged into an immense open plain, bordered by inaccessible mountains. Here the land had been cultivated as much for beauty as from necessity, for everywhere the useful was joined to the agreeable.'

Voltaire, *Candide,* or *Optimism,* 1759

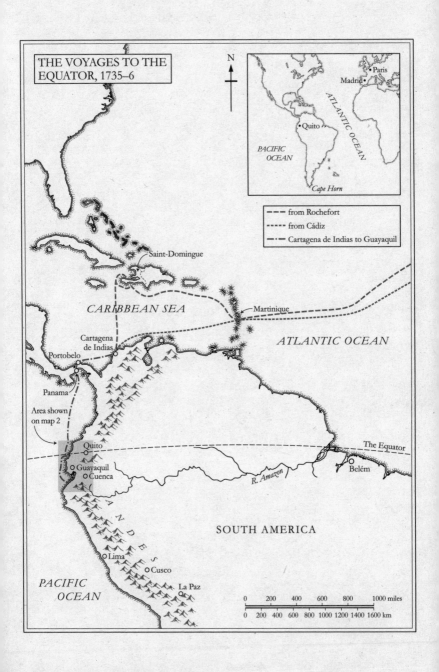

THE VOYAGES TO THE
EQUATOR, 1735–6

N

Paris
Madrid

ATLANTIC OCEAN

Quito

PACIFIC
OCEAN

Cape Horn

- - - - from Rochefort
········ from Cádiz
-·-·- Cartagena de Indias to Guayaquil

Saint-Domingue

CARIBBEAN SEA

Martinique

ATLANTIC OCEAN

Cartagena
de Indias

Portobelo

Panama

Area shown
on map 2

Quito

The Equator

Guayaquil
Cuenca

R. Amazon

Belém

A
N
D
E
S

SOUTH AMERICA

Lima

Cusco

PACIFIC
OCEAN

La Paz

| 0 | 200 | 400 | 600 | 800 | 1000 miles |

| 0 | 200 | 400 | 600 | 800 | 1000 | 1200 | 1400 | 1600 km |

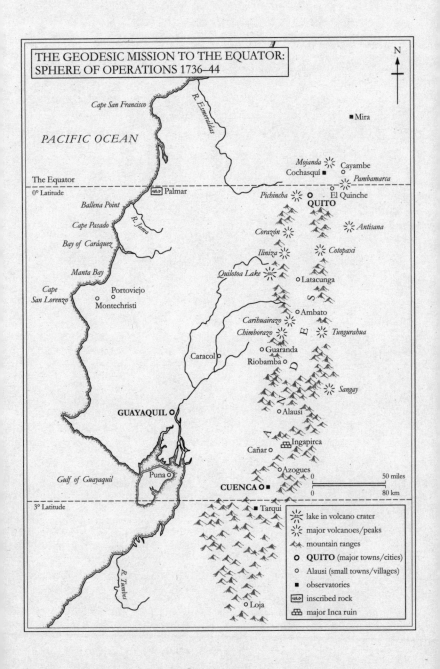

THE GEODESIC MISSION TO THE EQUATOR: SPHERE OF OPERATIONS 1736–44

N

Cape San Francisco

R. Esmeraldas

■ Mira

PACIFIC OCEAN

Mojanda ☀ ○ Cayambe
Cochasquí ■
☀ *Pambamarca*

The Equator
0° Latitude

⊡ Palmar

Pichincha ☀ ○ El Quinche
QUITO

Ballena Point

Cape Pasado

R. Jama

Corazón ☀

☀ *Antisana*

Bay of Caráquez

Iliniza ☀

☀ *Cotopaxi*

Manta Bay

Quilotoa Lake ☀

● Latacunga

Cape San Lorenzo

○ Portoviejo
○ Montechristi

A
N
D

● Ambato

Caribuairazo ☀
Chimborazo ☀

E
S

☀ *Tungurahua*

Caracol ○

○ Guaranda
Riobamba ●

☀ *Sangay*

GUAYAQUIL ○

○ Alausi

A

Ingapirca ⛬
Cañar ○

Gulf of Guayaquil

Puna ○

Azogues ○

0   50 miles
0   80 km

CUENCA ○ ■

3° Latitude

■ Tarqui

☀ lake in volcano crater

☀ major volcanoes/peaks

🝆 mountain ranges

○ **QUITO** (major towns/cities)

○ Alausi (small towns/villages)

■ observatories

⊡ inscribed rock

⛬ major Inca ruin

R. Tumbes

○ Loja

I

The tide was turning. Soon the moon would pull the sea down the black river. From the darkness came the gargle of water on wet mud and the eerie calls of unseen birds. On the west bank, the roofs and towers of Rochefort fused with the inky sky. A muffled *clunk* carried through the damp air. Boatmen were taking to their oars. The waiting was almost over. The ship's hull shivered as her three masts turned across the fading stars. Morning had broken on 12 May 1735.

Portefaix was crammed with over one hundred passengers and a cargo of grain and cannon. Like most ships of the French fleet bound for the colonies, the hold was loaded to capacity. All on board hoped for a safe passage. Eighteen years was a long time for a ship to be working the seas. The 117-foot keel had been laid in Toulon back in 1717 and the hull constructed from the recycled timbers of three decommissioned warships. With 22 eight-pounder cannon on the lower deck and 22 six-pounders on the upper deck, the ship was handled by a crew of 140 men and 5 officers. They were commanded by Lieutenant Guillaume de Meschin.

The last month had been frustrating. Included on the ship's manifest was the 'Geodesic Mission to the Equator', an unruly gaggle of *universitaires*, assistants and servants that had washed up on the Charente from all corners of France with an unimaginable quantity of baggage, scientific instruments and letters of authority from the king. They had more than twenty trunks of books. One of them had brought a dog. Instructions had been sent to Rochefort's navy intendant – the crown official charged with the port's operation – to supply the mission with swords, muskets, powder and ammunition, tents and blankets, surgical equipment and cooking utensils. More than sixty crates and trunks had accumulated on Rochefort's stone quay, together with an unpackable assortment of paraphernalia. The weight and bulk were too much for the ship. The mudbanks of the Charente and the shoals of the bay were notorious for snagging overloaded vessels. Confronted with the mission's excessive baggage, Lieutenant Meschin had been obliged to unload from the hold 140 barrels of grain. It had taken two days to rearrange the cargoes. This was achieved in the presence of Professor Bouguer, who – besides being a hydrographer and astronomer – happened to be France's leading expert on weight distribution in ships.

Meschin gave the order to weigh anchor and to warp the ship downriver on the ebb tide. By 10 a.m., they were moving along the sea-reach of the Charente towards the fort at the entrance to the estuary. *Portefaix* eased past the embrasures of Île Madame, into the open water of the bay. Then the wind died. Eyes turned to limp sails. As the anchor chain rattled to the holding ground off Île d'Aix, some of those on board wondered whether the false start was a bad omen.

Hours dragged into night. For four days, *Portefaix* rode at anchor. Then the wind returned and the sails were raised. Beneath taut arcs of canvas, crew and passengers watched the low shore of Île d'Oléron slide by the port rail until they were safely past the northern tip of the island and the cautionary finger of Colbert's lighthouse. Jean-Baptiste Colbert, *le Grand Colbert*, builder of lighthouses, roads, canals and of France, had laid the foundation stone of the French Academy of Sciences, the first learned society in France devoted to scientific research. Three of those on *Portefaix* were elected members of the Academy.

As *Portefaix* turned to the west, the deck began to heave then drop upon the ocean swell, and Professor Bouguer ejected the contents of his stomach. He had not wanted to join the mission. 'I had no intention of having anything to do with the enterprise,' he recalled later, claiming that the 'weak state' of his health had led to a 'repugnance' for sea-voyages. Pierre Bouguer was Royal Professor of Hydrography at Le Croisic, the key port on the Atlantic coast of Brittany, where he trained captains and pilots for a life at sea. But Bouguer was not a natural sailor and this was his first Atlantic crossing.

Untroubled by tempests, *Portefaix* sailed across the Bay of Biscay towards the tip of Spain. Off the rocky snare of Cape Finisterra, the crew and passengers saw the last of Europe. With every passing watch, they settled into the routines and rigours of life on board. Compared to the security of terrestrial France, a 650-ton ship carrying 250 people was claustrophobic and uncomfortable. Bouguer and the other two Academicians distracted themselves from nausea and tedium by learning how to use their new instruments.

They were sailing to South America in order to answer the outstanding question of the time: What was the true shape of the Earth? Most were agreed that Earth was not a perfect sphere. But was it elongated towards the poles, or flattened? Was Earth prolate or oblate? In the elongated camp were followers of the French philosopher René Descartes. In the flattened camp were followers of the English mathematician Isaac Newton. In the Parisian salons and cafés frequented by the Academy's elite, the Cartesians and Newtonians were fairly evenly split. Newton's theory was relatively recent and predicted that the centrifugal forces inside a fluid, rotating Earth were so great that it bulged at the equator and flattened at the poles.

It was more than an abstract debate. Without knowing the precise shape of the Earth, there could be no accurate maps or charts. The measurements provided by the Geodesic Mission to the Equator promised to make ocean navigation less dangerous and more profitable. Among those who grasped the geopolitical and economic benefits to France was the Minister of the Navy, Jean-Frédéric Philippe Phélypeaux, Count of Maurepas, who was masterminding a resurgence of French maritime power and prestige. Safe navigation was essential for the navy, and a French excursion into the Viceroyalty of Peru would be able to gather useful intelligence on the Spanish colonies of South America, with possible benefits to trading and political relationships between the two European superpowers.

On board *Portefaix*, the three Academicians bore the weight of expectation. Returning to France without a result was not an option. Key to resolving the prolate/oblate debate were the parallel lines of latitude that girdled the globe. The latitude of any point on Earth's surface was its angular distance from the equator. All places on the equator were therefore at zero degrees of latitude, whereas the North Pole was at 90 degrees north and the South Pole at 90 degrees south. By comparing the length of one degree of latitude in France with the length of one degree of latitude at the equator, it would be possible to discover whether Earth was prolate or oblate.

To calculate the length of one degree of latitude at the equator, the Academicians planned a two-stage process: firstly, they would lay out a virtual chain of triangles and use angular measurements to calculate the precise length of the chain. Secondly, they would use astronomical observations to fix the latitude at each end of the chain of triangles. By dividing the length of the chain of triangles on the ground (adjusted to sea level) by its length in astronomical degrees, they would be able to compute the length of one degree of latitude. Arriving at this figure was simpler on paper than it was in execution. The length of one degree of latitude was known to be around 60 miles but, to improve accuracy, they wanted to extend the survey to three degrees, so the total length of the chain of triangles would be nearly 200 miles. A geodesic survey of this type and scale had never been attempted in such difficult terrain: the region of equatorial South America selected for the survey was notorious for its rainforests, volcanoes and ravines, and for a medley of random hazards that included lethal diseases, dangerous beasts, appalling communications and suspicious Spanish officials. The mission was a very expensive, technically challenging, physically hazardous quest for a mathematical number. That number would be multiples of a French *toise*, a unit of measurement equivalent to 6 *pieds*. With the mission was an iron bar that had been forged and ground to exactly one *toise* by the specialist instrument-maker Claude Langlois. The bar would be used to calibrate all measurements during the expedition.

To their fellow passengers on *Portefaix*, the Geodesic Mission to the Equator looked an odd congregation. There were ten of them on board, attended by four servants. At their head were the three members of the French Academy. The eldest of the three was the Breton professor Pierre Bouguer, a man who had thought in numbers since boyhood in Le Croisic, where his father had been Royal Professor of Hydrography. At the age of sixteen, Pierre had inherited his father's post and, by the age of eighteen, he was a regular visitor to Paris and the Conseil de Marine, the Navy Council. He was geeky and pedantic, and the radical logic of his mathematical imagination was just what the navy needed. At twenty-three, he was invited to settle a debate within the French Academy of Sciences about two opposing methods of measuring the tonnage of a ship's hold. Six years later, the prodigal hydrographer won a prize from the Academy for a paper titled 'On the Best Manner of Forming and Distributing the Masts of Ships.' Among the research projects he had been working on when he received the call from Maurepas was a treatise on naval architecture aimed at replacing the unwritten trial-and-error habits of shipbuilding with mathematical rules founded on the laws of physics. The unfinished treatise was in his cabin on *Portefaix*. Bouguer's reluctance to join the mission had more to do with scientific ambition than with a fear of ocean voyages. A prolonged overseas sojourn would interrupt his studies. But Maurepas had wooed the aspirational, hard-up scientist with a consignment of valuable instruments and promises that his expenses would be covered and that he would be elevated within the Academy from *associé ordinaire* to *pensionnaire*.

Charles-Marie de La Condamine was three years younger than Bouguer and, like his fellow Academician, unmarried. As the other members of the mission would soon discover, La Condamine's character was founded on a disruptive combination of curiosity and recklessness. His background was conventional enough: a Parisian father who had been a collector of taxes and an education in the humanities and mathematics at the Jesuit college of Louis-le-Grand. But on completing his studies he had enlisted in the army and found himself on the front line, fighting Spain. In an episode that came to characterize his military career, he had climbed to a high point during the Siege of Rosas so that he could better observe the fall of enemy artillery shells, oblivious to the fact that his conspicuous purple cape was the reason that the rounds were exploding about his person. On his return to Paris, the war veteran fell in with the thinkers and doers of the Academy of Sciences and with the circle of intellectual reprobates that had been attracted to another ex-pupil of Louis-le-Grand, François-Marie Arouet, the outspoken historian and philosopher then writing under the nom de plume Voltaire. In 1730, La Condamine identified a loophole in government lottery regulations and formulated a plan with Voltaire and other friends to work the system. They all made a lot of money. That year, La Condamine became a member of the Academy, with funds. La Condamine and Pierre Bouguer – the reckless adventurer and the methodical mathematician – made an unlikely pair, but it was clear to many on board that they were destined to become close friends.

The third member of the Academy sailing on *Porte-faix* was a problem-in-waiting. Louis Godin was younger than Bouguer and La Condamine, better connected and the bearer of flaws that included arrogance and vanity. Born in Paris and educated in astronomy at the Royal College, he was accompanied by good looks, height and by a father who was a lawyer in parliament. Without publishing a paper, Louis had managed to become an *adjoint* member of the Academy at a mere twenty-one years of age. Once through its doors, he had sauntered its corridors of influence to edit the *Mémoires de l'Académie des Sciences*. In 1729, aged twenty-four, he married Rose Angélique Le Moine and the young couple – comfortably settled on the Left Bank near the Sorbonne – were soon parents of a son and a daughter. By 1730, Godin had managed to manoeuvre himself into a conspicuous role as the editor of *Connaissance des temps*, or *Knowledge of the Times*. This was the official astronomical ephemerides, an annually updated book of some 200 pages packed with tables and guidance on taking observations. It was the oldest and most revered publication of its type in the world and the editor's name appeared in capitals on the title page. Coinciding with Godin's editorship, a new section was added, listing the names and addresses of the *messieurs* who belonged to the Royal Academy of Sciences. A month after the 1734 edition went to press, Louis Godin had moved up the list from a lowly *adjoint* member to a *pensionnaire ordinaire* and had presented to the Academy a carefully prepared proposal to mount an expedition that would determine the shape of the Earth. Maurepas had put Godin in charge. But Louis Godin was the kind of man who could not run a *boulangerie*, let alone the world's first international scientific expedition.

Working closely with the three Academicians would be a quartet of specialists. The most experienced was Jean-Joseph Verguin, who would act as the mission's principal surveyor. A practised engineer trained in cartography and astronomy, Verguin would be responsible for producing the maps upon which the survey would depend. Helpfully, he was a trans-Atlantic veteran. Fifteen years earlier, he had sailed to the Caribbean and worked on a survey of Cartagena de Indias and on another of the Mississippi Delta. On his return to France, he had been employed in Toulon dockyard as an architect and draughtsman. In 1731, Verguin went to sea again, on a Mediterranean cruise that included the Greek islands and the North African coast. His large-scale maps of strategic ports such as Tripoli, and of key anchorages in the Ionian and Aegean seas, were models of practical clarity. At thirty-three, he was one of the older members of the mission. Verguin's maturity and experience would be important assets to the project. Like several other members of the expedition, he imagined that the geodesic survey might take him away for a couple of years at most, rewarded perhaps with a pension. His wife and two children would remain in Toulon while he was away.

The specialist tasked with recording in pictures the mission's progress was a New World innocent, Jean-Louis de Morainville. An artist and draughtsman in his late twenties, he left his wife in France so that he could work on the mission's drawings and maps.

The team's technician was a clockmaker called Théodore Hugo. His role would be the maintenance and adjustment of the mission's diverse inventory of instruments. These ranged from compasses and clocks to thermometers, barometers and the delicate pendulums that the Academicians intended to use for measuring the gravity of Earth, which – if Newton was correct – should be reduced at the equator because the bulge would place equatorial locations further from the planet's core. The workhorses for surveying on land would be a collection of quadrants for measuring angles. But the instrument most likely to keep Hugo awake at night was the huge twelve-foot zenith sector that would be needed for astronomical observations. Hugo was a skilled, versatile artisan, adept at working meticulously with metal, but his familiarity with astronomical instruments was limited.

Completing the tally of specialists was Joseph de Jussieu. His family were close with Godin, who admired the Jussieus for their successes in the world of botany and medicine. Joseph's older brother, Antoine, was director of the Jardin du Roi in Paris and his three-volume *Elements of Botany* had rewarded him with election to the Academy. Another brother, Bernard, followed his medical studies with a role in the Jardin du Roi, where he was developing a new method of plant classification. Their younger sibling, Joseph, had graduated with a doctorate in medicine and was teaching at the University of Paris when he received the call to South America. Introspective and vulnerable, Joseph de Jussieu was expected to function as expedition doctor and botanist.

The three remaining French members of the mission were a mixed bag of mates and favours. Jean-Baptiste Godin des Odonais was Louis Godin's first cousin. Just into his twenties, Jean-Baptiste had been set to spend his life at a loose end until he was plucked from the family estate on the languid banks of the River Cher in the rural heartland of France. He had no geodesic experience to contribute, and joined the mission as a general assistant.

Jacques Couplet-Viguier was the nephew of one of Louis Godin's friends, Nicolas Couplet de Tartreaux, treasurer at the Academy. Jacques, too, was an expedition novice. At seventeen, he was the youngest member of the mission, but he boarded *Portefaix* buoyed by the achievements of his ancestors. His grandfather had taken part in Cassini's surveys of France and his uncle, Nicolas, had spent time in South America collecting astronomical observations. Like Godin de Odonais, he would function as a general assistant.

Jean Seniergues was a close friend of Joseph de Jussieu. Both had just turned thirty when *Portefaix* sailed. Seniergues was unabashed by his motives for accepting a berth on *Portefaix*. Gold mines and private medicine would make him rich. He was single and had everything to gain. By profession, he was a surgeon and therefore ranked lower than his medically qualified friend Joseph. But his experience with treatments and pain relief were expected to be an asset on the equator.

Included on the mission's French passport were four unnamed servants, a shadowy quartet whose identity and accomplishments would be virtually excluded from the published memoirs and reports. Over the coming years, the mission's servants would be joined or replaced by other *domestiques* and a changing cast of local guides, porters, boatmen, muleteers and labourers, without whom the mission would have foundered.

Not on *Portefaix* were two key members of the team. To gain access to the Viceroyalty of Peru, Maurepas had convinced Spain's Minister of the Navy – José Patiño – that the French mission would be of enormous benefit to Spanish navigation. To encourage Spain's cooperation, Maurepas had offered the mission's services in measuring the latitude and longitude of key locations on the coast of Peru. Patiño consulted his king and the Council of the Indies, and the response was positive, with the caveat that the mission must include 'two intelligent Spaniards'.

For more than a month, *Portefaix* rolled and pitched on the Atlantic Ocean. The first cracks began to appear in the team. One of Louis Godin's more divisive characteristics was his habit of looking down upon his fellow travellers. In particular, this rankled with the surgeon, Jean Seniergues, who claimed later that Godin had 'wished to erect himself as the Grand Master'. Others on the mission felt the same way. The discord was compounded by an early alliance between La Condamine and Bouguer, whose doubts about Godin's leadership were accumulating with every day spent on *Portefaix*.

On 20 June, frigate birds were spotted flying low over the waves in search of squid and fish. Then, on the 22nd, as the sun was beginning to tint the eastern horizon, a mountain solidified in the sea mist. Martinique. The relief among the passengers was palpable. They had survived the Atlantic. The onward voyage should be little more than island hopping. Lieutenant Meschin took *Portefaix* around the tip of the island to its west coast and sailed his ship past the headland known as Pointe des Nègres into Fort-Royal Bay. The anchor fell before a formidable array of cannon muzzles protruding from the waterline fort. They stayed for ten days while cargo was unloaded and essential supplies taken on board.

For the scientists, it was an opportunity to rediscover

their land-legs and explore. Martinique was a defiled garden, an island known by native Caribs for its flowers but snatched in 1635 by the French Compagnie des Îles d'Amérique. The Caribs were exterminated and by the time *Portefaix* was swinging at the sheltered anchorage beside Fort-Royal, the island's most accessible slopes were covered in sugar and coffee plantations and Martinique had become a Caribbean prison for 60,000 black slaves.

The three Academicians went romping with their instruments, climbing Mount Pelée to calculate its altitude and fixing the latitude and longitude of Fort-Royal. Engineer Verguin found his feet by drawing an accurate chart of the bay, complete with depth soundings and the location of dangerous shoals. For Joseph de Jussieu, Martinique was a botanical wonderland. Sticky with sweat, he tramped the slopes in search of exotic plants and fruits. Already he had learned much about the island from his brother Antoine, whose professorial role at the Jardin du Roi made him a convenient source of botanical information. 'I believe I shall become quite at ease in the tropical climate,' Joseph wrote excitedly to his brother in Paris. Jussieu selected plants to send back to France on the next available ship. All this scientific scampering was, in a sense, superfluous, because the Jardin du Roi already had the plants, and the location of Fort-Royal and the altitude of Pelée were known, but for Godin, La Condamine, Bouguer, Verguin and Jussieu it was all good practice for the fieldwork they would soon be undertaking in far more exacting terrain.

Two days before *Portefaix* sailed, there was an unsettling

death on board. Among the passengers who joined the ship at Fort-Royal was a Swiss sergeant. He was, noted La Condamine, 'a robust man'. But he was 'carried away in less than a day by *maladie de Siam*, so common in our islands'. Siam disease, or 'Black Vomit' – after the internal bleeding that invariably flooded the digestive tract – was thought at the time to have been imported to the Caribbean by a ship bearing French settlers. Soon, it would become known as yellow fever, a virus spread by mosquitoes. Symptoms were a flu-like combination of headaches, fever and muscle pain, followed by nausea and vomiting, jaundice, bleeding, seizures and organ failure. Survival rates were not more than 50 per cent. The disease was common in South America. From Martinique on, every member of the mission would be one mosquito bite from mortality.

There was no reason for the departure of *Portefaix* from Martinique to be complicated but, for the second time, the French scientists succeeded in frustrating the schedules of Lieutenant Meschin. First, Godin tried to extract *livres* from the intendant of Martinique, ostensibly to cover the mission's expenses while they had been on the island. And then La Condamine went down with 'a violent fever'. It took hold of his body with ferocious speed. Later, he related that the symptoms 'made it appear that [he] was attacked by the same disease' as the Swiss sergeant. Yellow-eyed, feverish and aching, La Condamine was in no condition for another sea voyage. Faced with being left behind on Martinique, he agreed to be drained of blood and flushed through with emetics.

Bled and purged, the Academician was loaded on board and, on the evening of 4 July, *Portefaix* sailed away from Fort-Royal.

The passage from Martinique should have been straightforward. In good winds, French ships could follow a north-westerly course across the Caribbean to the colony of Saint-Domingue, which occupied the western part of the island of Hispaniola. It was here that the mission was due to disembark from *Portefaix* and secure berths and hold-space on a ship that could take them onward to the Spanish port of Cartagena de Indias, on the mainland of South America. Four days out of Fort-Royal, *Portefaix* sailed into a thick bank of fog.

2

As *Portefaix* dissolved into fog off Hispaniola, a sleek pair of Spanish warships received a nine-cannon salute off Cartagena de Indias, the portal to South America. *Nuevo Conquistador* and *Incendio* had sailed from Cádiz on 26 May, some fourteen days after *Portefaix* had left Rochefort. Compared to the meandering progress of the aged French warship, the Spanish men-of-war had crossed the Atlantic like thoroughbreds. On board were the 'two intelligent Spaniards' detailed to accompany the Geodesic Mission to the Equator.

Jorge Juan y Santacilia and Antonio de Ulloa y de la Torre-Guiral were both graduates of the Academy of Navy Guards, the elite corps of young men picked from Spanish nobility to be trained in mathematics, astronomy, navigation and associated subjects ranging from trigonometry and hydrography to cartography and firearms. Both men had seen active service. Jorge Juan had sailed with the Spanish Mediterranean fleet against roving corsairs and had been with the squadrons that took Oran from the Ottomans. Ulloa had sailed on a two-year voyage with a fleet of galleons across the Atlantic, calling at key Spanish ports around the Caribbean. Selection for the geodesic mission promised instant promotion. Their Navy Guard commanders had been told to choose 'two persons, whose dispositions not only promised a perfect harmony and correspondence with the French academicians, but who were capable of making, equally with them, the experiments and operations that might be necessary in the course of the enterprise'. In order to invest the two Navy Guards with sufficient authority during their mission, both had been promoted to the rank of lieutenant. Jorge Juan was twenty-two and Ulloa was nineteen.

As the two lieutenants watched the muzzle smoke drift across Cartagena Bay, they had reason to wonder. An expedition to the Andes was not a normal mission for members of the Navy Guards. They had been trained for oceans, not mountains. If this strange commission was to bring rewards for Spain – and for them – they would have to work well together. It was perhaps to their advantage that they had such dissimilar backgrounds. Jorge Juan had been born on the Alicante coast and was three when his father died, leaving two uncles to oversee his Jesuit education and then – when he was twelve – send him across the water to Malta, where he had been received into the Order of St John of Jerusalem. At fourteen, he had been granted the title of Commander of Aliaga in Aragón and he left the island as a Knight of the Order of Malta, an office that demanded life-long celibacy. Jorge Juan was a strongly built man of medium height and was remembered by one who knew him well as having 'a pleasant and placid countenance'. He 'ate frugally', his habits 'were those of a Christian philosopher' and he 'did not judge men in view of the place from where they came'. He had high expectations, particularly of himself. His nickname at the Guards Academy had been Euclid.

Jorge Juan and Ulloa had separated for the voyage, a practice they would repeat whenever possible. If one of them were lost, the other would continue with their mission. As the older and more senior by training, Jorge Juan sailed on the larger of the two ships, *Nuevo Conquistador*. The 64-gun man-of-war had been tasked with conveying to Cartagena de Indias the new Viceroy of Peru, José Antonio de Mendoza Caamaño y Sotomayor, Marquis of Villagarcía de Arousa. The relationship that Jorge Juan forged with Villagarcía during the voyage was to prove crucial.

While Jorge Juan supped with the Viceroy, Ulloa was being pitched about on the confined deck of 50-gun *Incendio*. The eager nineteen-year-old occupied himself by keeping a log of the variations in magnetic declination: the angle between true north and magnetic north. Ulloa's aristocratic father, Bernardo de Ulloa, was a published economist and young Antonio had grown up in the buzzing international city of Seville – the port on the Guadalquivir that, 200 years earlier, had welcomed home the survivors of Magellan's first circumnavigation of the world. At the age of thirteen, he had been packed off to the Academy of Navy Guards in Cádiz, but delayed enrolment for two years while he took his self-funded Caribbean cruise. Returning in 1732, he had been admitted to the Navy Guards in November of the following year, just in time to sail with a fleet to reinforce Naples. When he returned to Spain, late in 1734, preparations for the Geodesic Mission to the Equator were already advanced. Although he had received less formal training than his older compatriot, Ulloa's trans-Atlantic excursion to the Caribbean had prepared him well for the scientific epic then being planned.

The orders handed to the two lieutenants by the Minister of the Navy, Patiño, had extended beyond working with the French to measure one degree of latitude. The Spanish had some private projects on the side. Jorge Juan and Ulloa had been instructed to undertake surveys of colonial ports and territories. It was a brief that included the fixing of locations through astronomical observations, map-making, natural history, urban and cultural geography and the disposition of defence. In particular, they were to report on corruption within Spain's colonial administrations. Much of this material was to accumulate in a 'secret dossier' unknown to the French Academicians or to colonial administrators. Jorge 'Euclid' Juan would take the lead and focus on the advanced observations and mathematics, while Ulloa would cover the map-making and more descriptive tasks.

On landing at Cartagena de Indias, Jorge Juan and Ulloa learned from the governor that 'the French academicians were not yet arrived, nor was there any advice of them'. This was a disquieting setback. Somewhere between France and the Caribbean, the ship bearing the Frenchmen had disappeared. 'Upon this information,' remembered Ulloa, 'and being by our instructions obliged to wait for them, we agreed to make the best use of our time.'

For Ulloa in particular, this was a fruitful opportunity to test his investigative skills. He had visited Cartagena de Indias on his teenage adventure with the galleons and would have been familiar with – and perhaps even met – the city's famous military engineer, Juan Herrera y Sotomayor. Sadly for the two young lieutenants, Brigadier Herrera had died three years earlier, but his legacy could be seen across the city and bay in the defensive forts, bastions and batteries he had commissioned. Herrera had come to America in 1681 and over fifty years had worked his way up from a humble lieutenant in the garrison of Buenos Aires prison to the engineer who repaired and improved the defences of Panama, Portobelo and Cartagena de Indias, where his innovations included Dutch-style lock gates to improve river navigation on the Magdalena and the creation of the first engineering school in the New World: the Academia de Matemáticas y Práctica de Fortificación.

The two young lieutenants leapt into action. While waiting for news of the French, they would undertake a comprehensive survey of Spain's American hub. Cartagena de Indias had been founded back in 1533 by a Spanish conquistador, Pedro de Heredia, who identified the bay as a strategic anchorage. The natives whose villages dotted the shore were killed or driven inland. As the best natural harbour on the north coast of South America, Cartagena became an immediate target for nations competing for control of their respective 'New Worlds'. Eleven years after its founding, Cartagena was ransacked and ransomed by the French. In 1585, Sir

Francis Drake and his fractious band of heretical English pirates attacked, burned and again ransomed the city. By the 1730s, Cartagena had been secured with forts and gun batteries, but with the British lurking again beyond the horizon, it was no time for complacency.

There was one impediment to their impromptu survey. Before leaving Spain, Jorge Juan and Ulloa had ordered the most up-to-date instruments from Paris and London, but these had not arrived by the sailing date of *Nuevo Conquistador* and *Incendio*. Without instruments, there would be no surveying or observations. Happily, they learned through the governor that Brigadier Herrera's instruments were still in the city, along with some of his plans and maps. It was a very lucky break. Ulloa began to make the 'necessary additions and alterations' to Herrera's earlier work. It was a challenging task. Cartagena and its bay covered an area of over a hundred square miles. Day after day, Jorge Juan and Ulloa measured and observed. Blanks were filled on Herrera's maps and locations adjusted. The final clean copy in pen and ink carried the names of Ulloa and Herrera, framed within an exotic cartouche showing two native warriors with bow, quivers and spear leaning casually on the lettering of the title, which included the revised latitude and longitude for the city. The map was about the size of a small table. As an aid to the user, and as a guide while compiling the map, Ulloa had drawn a faint grid of squares with 2,000 *pieds de Rhin* – Rhine feet – between the lines. The overall scale worked out at around 1:25,000, so it was detailed enough to show features such as buildings. On land, he had marked prominent hills, vegetation, rivers, roads and the outlines in red of Spanish fortifications and gun batteries. The city itself appeared in pink, rimmed with walls and redoubts. Emphasizing its dual role as a land map and a sea chart, Ulloa had carefully marked with pecked lines the presence of shallow water, while small numerals dotted across the bay represented the depth in fathoms of places where soundings had been taken. The completed map was both a memorial to the great Herrera and a validation of nineteen-year-old Ulloa. But it was for Spanish eyes only. In the wrong hands, the new *Plano de la Cyudad y Bahya de Cartagena de las Yndias* was a very detailed invasion map.

For three days, *Portefaix* warily coasted the shrouded shores of Saint-Domingue. Then the fog lifted and Lieutenant Meschin was able to bring his ship into the anchorage at Fort Saint Louis on the southern side of the island. As the most important French colony in the Caribbean, Saint-Domingue would provide all the remaining necessities for the mission. It was the last chance to stock up on French territory and it was the setting for the mission's final severance from home. The Governor-General of Saint-Domingue, the Marquis de Fayet, had been instructed to do all in his power to ease the onward passage of the Academicians and their entourage. The Academicians did their best to make his life as difficult as possible.

Instead of staying with the ship and the mission, Godin and La Condamine decided to hike across the island and rendezvous with *Portefaix* at the port of Petit-Goâve, the colonial capital of Saint-Domingue. It was an adventurous initiative and it did have an element of scientific reasoning, for the two Academicians intended to use the trek as an opportunity to take astronomical observations. The straight-line overland distance from coast to coast was a mere 20 miles, but it was eight days before Godin and La Condamine reappeared at Petit-Goâve. A more experienced leader than Godin might have chosen to keep his expedition together.

At Petit-Goâve, the mission had to bid farewell to *Portefaix*. Ideally, the ship would have continued on to Cartagena de Indias, but *Portefaix* was scheduled to sail on 11 August for Louisbourg. Despite the urgings of Maurepas in Paris, Fayet had been unable to find an alternative ship large enough to accommodate the mission and its baggage. There were plenty of vessels at the Petit-Goâve anchorage, but all were too small. The crates, trunks and loose luggage were laboriously carried from *Portefaix* to the quay.

For three months, the mission was marooned with an unlimited expense account on Saint-Domingue. Later, Fayet would complain by letter to Maurepas in Paris that the presence of the mission had cost the colony around 150,000 livres. There were worse places to be stranded. The expedition leader dedicated himself to fieldwork in a brothel run by Bastienne Josèphe, a freed slave. One of her staff – known only by the name Guzan – formed a particular attachment to Godin, who insisted that the mission's draughtsman, Morainville, turn his hand to portraiture and produce likenesses of both Guzan and Bastienne. Before leaving the island, Godin blew 3,000 livres of mission funds on a diamond for Guzan, an extravagance that prompted the melancholy doctor Jussieu to observe that the mission's leader had 'for some time set astronomy aside to take care of more pressing affairs'.

Between expensive bonks in Chez Josèphe, Louis Godin experimented with his pendulum. He was joined by La Condamine and Bouguer. Reports were sent back to Paris for publication, reassuring the Academy that its scientists were engaged in valid endeavours. Jussieu, unsettled by the antics of his leader and anxious about his role on the expedition, busied himself on the island collecting seeds and recording species. As he had in Martinique, Verguin compiled a map. Fevers struck several members of the team. Bouguer's servant died. He was latitude's first martyr. La Condamine recorded that the loss 'was amply compensated by the Negro slaves made available to us at the King's expense'. Petit-Goâve's population included 2,000 slaves. Godin, Bouguer and La Condamine selected and bought three men to accompany them to South America. Bouguer chose a new servant named Grangier.

While humidity, distractions and insects slowed the pace on Saint-Domingue, events were moving quickly in Paris, where a charismatic Academician – and friend of La Condamine – called Pierre-Louis Moreau de Maupertuis was devising an expedition that threatened to shove the Geodesic Mission to the Equator into the shade. Maupertuis was one of the Academy's star scientists.

With Godin, René Antoine Ferchault de Réaumur and Jacques Cassini, he was one of the senior *pensionnaires ordinaires*, a Newtonian mathematician whose interest in the shape of the Earth was more than academic. Inspired in large part by the departure of the Geodesic Mission to the Equator, Maupertuis had succeeded in convincing his Newtonian allies within the Academy to support a second French geodesic mission. Led by Maupertuis himself, its goal would be to measure the length of one degree of latitude as close to the North Pole as possible. The location Maupertuis had in mind was a long river valley northwards from the Gulf of Bothnia, through the forests of Lapland to the Arctic Circle. With a latitude figure on the Arctic Circle and another at the equator, Earth would have been measured close to its physical extremities and, argued Maupertuis, 'France shall assuredly be doing the greatest thing ever for science.' The Arctic Circle was so much more convenient to France than the equator that Maupertuis was virtually certain to be back in the Louvre with a figure for the shape of the Earth before Godin's expedition had completed their measurements in South America. The Minister for the Navy, Maurepas, backed the Arctic Circle expedition, and by the beginning of September the Academy learned that Louis XV had given his royal consent. On 8 September, while Godin's mission was frittering time in Saint-Domingue, Maupertuis picked up his pen and began writing to La Condamine: 'You'll perhaps be surprised when you know that there is going to be a voyage to the North.' It would take over a year for the letter to catch up with the mission. In the meantime, nobody on Saint-Domingue knew that the Academy and the French government had launched a second expedition.

On 30 September, a two-masted brigantine sailed into Petit-Goâve and the onward route to South America opened up for the French scientists.

Vautour had sailed from Rochefort. She was far smaller than *Portefaix*, with a dozen guns and a limited capacity for passengers and cargo. But her commander, Lieutenant Louis du Trousset, Count of Héricourt, agreed to take the mission on to Cartagena de Indias. Another four weeks were consumed while the expedition collected and re-packed its crates and trunks, and assembled provisions needed in South America. Among the essentials sourced on Saint-Domingue were the 'field tents' that would provide shelter during the geodesic survey. The intendant at Rochefort had provided the expedition with three *canonnières*, crude, round-ended military tents formed by throwing a shaped canvas over a ridge pole. But he had also equipped the mission with a much larger, officers' *marquise*, a rectangular tent with high sidewalls and a secondary flysheet to provide additional weather protection. Godin had appropriated the officers' *marquise*, leaving La Condamine and Bouguer with the prospect of grovelling damply in the squaddies' *canonnières*. Spotting an inconvenience that could be resolved at government expense, La Condamine took Godin's *marquise* to a workshop in Petit-Goâve and had it used as a template for the construction of two large, double-skinned tents, one for himself and one for Bouguer. More small tents were bought. Meanwhile, Héricourt was busy concealing contraband in the ship's lower holds.

The smuggling was more than a bit of private enterprise by a dodgy mariner. Héricourt's illicit merchandise was loaded with the complicity of Saint-Domingue's Governor-General, Fayet, who was responding to a directive from Maurepas 'to try to do some commerce with the Spanish, or to lay the foundations . . .' Maurepas had told Fayet that an *'homme de tête'* – a head man, a boss – should be put in charge of the operation. It was a high-risk deceit. The year before, Maurepas had provided guarantees to Spain's Minister of the Navy, José Patiño, that France would not attempt to subvert Spanish trade routes by smuggling merchandise into America. If the true contents of *Vautour*'s hold were discovered by Spanish officials, the mission might be wrecked.

On 21 October, Héricourt ordered his crew to the halyards and set the helm of *Vautour* for Cartagena de Indias, 300 miles away on the far side of the Caribbean Sea.

Most of the mission's members had grown used to the luxuries of terra firma and it was a long two weeks at sea. But for La Condamine – not a man to squander an episode of adversity – there was unexpected relief in the tedium. Serving as second officer on *Vautour* was an aspiring poet he knew from Paris. Eighteen months earlier, La Condamine had dined with Jean-Baptiste Sinetti at the house of Voltaire. The chance reappearance of the poet played to the romance of their American odyssey. Effusively, he wrote to Voltaire:

Guess who's the second officer of the King's vessel, armed in St Domingue to take us to the Spanish coast . . . It is Monsieur Sinetti, with his fat jowls, none other. When you found it bad enough that he went to the islands, you never suspected he was predestined to be Jason of the modern Argonauts.

La Condamine portrayed the passage to Cartagena de Indias as a monotonous cruise enlivened by the plays and poems of their distinguished Parisian friend: 'Quoting extracts from *La Henriade*, *Zaïre* and *Adélaïde* was,' wrote La Condamine, 'the only way of bringing charm to our boredom.'

On the morning of 16 November, word reached Cartagena de Indias that 'a French armed vessel' had anchored during the night on the far side of the bay, under the guns of Boca Chica. Jorge Juan and Ulloa were rowed out to *Vautour* to meet 'the long-expected gentlemen'.

For both contingents, it was a surprising encounter. As Jorge Juan and Ulloa were introduced to the French scientists, they discovered that five men whose names they had been given had doubled to ten, plus servants and slaves. And three of the five were not on the ship. Only the names Godin and La Condamine tallied with the original list. Outnumbered by so many from France, the two Spanish lieutenants would need to comport themselves with caution. La Condamine was underwhelmed by his new colleagues: 'This is how we begin in Spain,' he wrote to Voltaire, 'with men who wear their love of Physics as trinkets.' Jussieu was kinder. The sensitive doctor found Ulloa and Jorge Juan 'friendly gentlemen with extremely sweet characters, very sociable, well born, very knowledgeable in mathematics and who speak French to make themselves easily understood'.

After waiting so long for the French to arrive, Jorge Juan and Ulloa were quietly impatient to set off for Quito and the equator. But the mission had swollen to a size unmanageable by a single leader. The ten Frenchmen and two Spanish lieutenants were accompanied by no less than fourteen *domestiques*. For eight days, this muddled agglomeration milled about in Cartagena de Indias.

There were a couple of outstanding issues preventing the mission from proceeding into the wilds of South America. The first was a chronic shortage of funds. The long stay on Saint-Domingue, and Godin's promiscuous spending, had whittled the mission's funds down to a mere 9,000 livres. Godin met with the Cartagena de Indias correspondent of the French banking firm Casaubon, Béhic and Company, who ran one of the most successful silver-smuggling operations in Cádiz and with whom Maurepas had set up a line of credit worth 4,000 pesos. For the time being, Godin had secured sufficient funds to keep the mission moving. Jorge Juan and Ulloa were reassured by the arrival of their back pay. The other issue was geographical:

'Our intention being to go to the equator with all possible expedition,' wrote Ulloa, 'nothing remained but to fix on the most convenient and expeditious route to Quito.' The city the mission was trying to reach lay far to the south, along the Andes, an arduous and sometimes dangerous overland trek of some 400 leagues – 1,600 miles – reckoned to take at least four months. The alternative was a complicated detour by sea and land, from Cartagena de Indias along the Caribbean coast to the anchorage of Portobelo – where the Panamanian isthmus was at its most narrow – and then a short overland trek to the city of Panama, on the Pacific, where they would have to pick up a passing ship that could convey them south down the coast to Guayaquil, where they could disembark and then travel by riverboat and mule through rainforest and mountains to Quito, over one hundred miles inland. As they debated options, the wet season had already started. The trails through the Andes would be deep in mud and the rivers swollen with run-off.

La Condamine was one of those arguing against the overland route. Expanded to an unruly crowd of twenty-six men, most of whom were physically unfit after months on ships and in ports, the mission would have to be accompanied by at least as many guides, porters and muleteers. Fifty or so men trekking with around one hundred loaded mules through the Andes would move very slowly and chaotically. And as La Condamine pointed out, the baggage that had been crated and boxed for conveyance in the hold of a ship would have to be unpacked and redistributed as mule loads. The large, fragile instruments that had been shipped out in their assembled form would have to be dismantled. Pushing for the long overland route were unidentified 'interested counsels' who La Condamine declined to identify in his memoirs. La Condamine had another reason for taking the Pacific route. With Bouguer, he was angling for the mission to take advantage of the difficulties posed by the wet season in the Andes by pausing during their sea voyage to Guayaquil and undertaking fieldwork and observations where the equator crossed the Pacific coast. An equatorial reconnaissance would also allow them to investigate the possibility of an alternative, coastal venue for the 200-mile chain of geodesic triangles. La Condamine and Bouguer won the argument. The decision was taken to stay with *Vautour* and to sail for the Panamanian isthmus.

Final preparations were made for departure. *Vautour* took on provisions and water. For security during the short voyage to Portobelo, the twenty-strong detachment of Swiss soldiers from the garrison at Petit-Goâve would stay with the ship.

On 24 November, the Geodesic Mission to the Equator gathered for their last passage by ship in Atlantic waters. The story of science was about to turn a new page. The full complement of the world's first international scientific expedition was now assembled. If the twelve men from two countries that gathered that day on the deck of a salt-streaked brigantine in the great bay of Cartagena de Indias had been subjected to psychological profiles, they would have been deemed an utterly dysfunctional team. The Disparate Dozen of the Enlightenment were about to embark upon an extraordinary adventure.

3

Vautour's capstan groaned in the morning breeze as the anchor broke the surface and thudded against the timber bow. The helmsman turned the ship between the shoals towards the narrow neck of Boca Chica and the mute barrels of the gun batteries. Sheets were tightened and sails trimmed. The deck began to rise and fall with the swell of open sea. Pierre Bouguer tried not to think about his stomach.

The passage west from Cartagena de Indias should have been a straightforward run along the coast towards Portobelo, where the mission intended to disembark for its traverse of the Panamanian isthmus to the Pacific Ocean. But the weather was unkind and the Caribbean was heaving with gales churning in from the northeast. For five days, *Vautour* crashed through high seas.

At five in the evening on the 29th, the welcome outline of the headland known as 'Ship Point' was spotted on the horizon. But with the wind now from the south, the crew of *Vautour* were forced to tack the vessel to and fro towards the narrow gap in the coast. Then the sails sagged and the wake turned to glass. Obtusely, an offshore wind sprang up. The ship's boat was launched and the crewmen bent over their oars to tow *Vautour* towards the looming towers of Castle Todo Fierro – the Iron

Castle – guarding the entrance to the harbour. With the anchor holding, baggage and instruments were stacked on the deck and the mission prepared to be ferried ashore. After the delays at Saint-Domingue and Cartagena de Indias, it was essential not to waste more time and money at another port. Besides, Portobelo was a dump.

Back in 1502, Christopher Columbus had chosen this deep, sheltered sea-inlet as a convenient anchorage. He named it Porto Bello: fine harbour. It was a hellhole of the lowest order. Excessive heat, violent rainstorms and a thick ceiling of trees conspired to create a sticky microclimate seething with biting insects. Entire ships' crews were known to have succumbed to disease. *Vautour*'s French sailors knew it as the *Tombeau des Espagnols*. It was a Portobelo precaution to send pregnant women away to the relative safety of Panama on the west coast of the isthmus. For the slaves charged with loading and unloading ships and barges, and with hauling overloaded wooden sledges along the town's mud-logged ways, the humidity was hideous. The custom of rehydrating with brandy increased the death toll. Nine years earlier, an attempt by a British fleet to intercept Spanish treasure at Portobelo's periodic trade fair had to be aborted after the British lost half their men to disease. To the mission's doctor, Jussieu, Portobelo was 'the most unseemly and unhealthy place in the universe'. But Portobelo was also the transshipment point for people and freight crossing the narrow neck of land dividing the Atlantic from the Pacific. A stopover here was unavoidable.

Portobelo to Panama was a straight-line distance of only 40 miles or so but the overland trail between the two places was – La Condamine had learned – 'one of the worst in the world'. Less difficult would be the alternative route by riverboat up the River Chagres, which would reduce the walking to a single day over the final watershed to the Pacific. The crates of scientific instruments, the 21 trunks of books, the 9 barrels of French spirits, the 225 pounds of gunpowder, the 28 tents, the dress clothes and wig powder, the stashes of 'Andalusian tobacco . . . and other small items' that had been logged by the governor of Portobelo while he checked for smuggled goods, would have to be prepared for carriage by shallow-draught *chata* and then by porters. Godin sent an urgent request for help to the provincial administrator, the president of the *audiencia* of Panama, Dionisio Martinez de la Vega, enclosing the orders from King Philip V of Spain. With Jorge Juan and Ulloa to verify the Spanish interest in the mission, the president's prompt response was – according to Ulloa – 'not in the least wanting'. He would send boats.

Meanwhile, the mission was stuck in Portobelo. Day by day, money and time slipped away. Jussieu, who had been ill during the rough voyage from Cartagena, managed to cure himself, 'proof', observed La Condamine, 'of his art by re-establishing himself in a place where Spanish fleets often lose a third and sometimes half of their crews'. But the draughtsman, Morainville, fell sick. Both La Condamine and Ulloa were stung by scorpions.

In trying circumstances, the leading members of the

mission tried to fulfil their roles as visiting scientists. Godin and Bouguer set up the pendulum against a wall in their lodgings in order to measure Earth's gravity (La Condamine complained that his absence from these observations was due to his lodgings being worse than those of the other two Academicians). Verguin and Ulloa knew Portobelo from previous voyages and managed to use their time well: Verguin gained approval from the port's governor to compile a map of the port and its defences, while Ulloa used the enforced stopover to fill his notebook with information of interest to the Spanish crown. With Jorge Juan, he observed and noted the harbour's dimensions, tides, winds, anchor holdings ('clayey mud, mixed with chalk and sand') and its defences. They took observations of the Pole Star and of the angle between the meridian and the sun – its azimuth – and determined the magnetic declination of the compass at 8° 4′ easterly. They took temperature readings with the Réaumur thermometer and described how to forecast local tempests by observing the density and movement of cloud on the peak at the harbour entrance. They consulted 'some intelligent persons' about the stories that Portobelo's appalling climate caused imported hens to cease laying and cattle from Panama to become inedible through weight loss. And they amassed notes on the wildlife that roamed the port: 'tigers' that came down from the mountain forests at night to snatch livestock and small boys; the giant toads that appeared after rain, covering the mud so completely that it was impossible not to tread on them; the deadly snakes. Portobelo's weirdest creature was 'nimble Peter', an ironic nickname for an animal of 'extreme sluggishness'. Ulloa described it as a 'middling monkey . . . of a wretched appearance' whose occasional movements were accompanied by 'such a plaintive, and at the same time so disagreeable a cry, as at once produces pity and disgust.' The lieutenant had met a brown-throated sloth.

Three weeks after the mission arrived in Portobelo, twenty black slaves rowed a *chata* into the anchorage. A second *chata* followed. The Europeans could not leave fast enough: 'Immediately on their arrival,' recorded Ulloa, 'we put on board the instruments and baggage, belonging both to the French gentlemen, and ourselves; and on 22nd of December 1735, departed from Porto Bello.'

It was a rough passage along the coast. In each boat, the passengers occupied a makeshift cabin at the stern: 'a kind of awning supported with a wooden stanchion reaching to the head', remembered Ulloa. Baggage was protected by hides from sea spray and rain. An onshore wind meant that the two heavily laden *chatas* had to be rowed out of Portobelo's harbour, the backs of the oarsmen bending in time to the shouts of the pilot. By nine that morning, they had cleared the headland and were under sail, fighting a 'fresh gale'. Slowly, the unwieldy craft butted west through the whitecaps, and by four in the afternoon, they were safely rowing into the mouth of the Chagres beneath the guns of San Lorenzo fort. Across the river from the fort, the pilots steered the two *chatas* on to the sand in front of the customs house, where they spent the night. Next morning, they began rowing upriver.

Since leaving the shores of Europe, the vessels had been getting smaller. They had voyaged from ocean to sea to estuary and now river. The Chagres snaked inland, a dark river walled by dark forest. Alligators lurked where liquid met solid. La Condamine and Verguin worked on a map of the river. Ulloa scribbled in his notebook, trying to make sense of what he was seeing and hearing: 'The most fertile imagination of a painter can never equal the magnificence of the rural landscapes here drawn by the pencil of Nature . . . monkeys, skipping in troops from tree to tree . . . the wild and royal peacock, the turtle-dove, and the heron . . . the pine-apples, for beauty, size, flavour, and fragrancy, excel those of all other countries'. With every mile, the river narrowed and the current strengthened. Labouring in hot, humid sunlight over heavy timber oars, the rowers weakened. On the 24th, they found the effort too much and resorted to quanting the two boats forward using long poles. Obstructions were many. Toppled trees of unimaginable girth reached across the navigable channel, ready to cause a capsize. A tremble in the suffocating heat warned of rapids, where each *chata* would be lightened while the slaves hauled the timber craft up a staircase of tumbling water. For another three days, they pushed on as the Chagres unravelled in serpentine loops. At eleven in the morning on the 27th, they reached a muddy river-port called Cruces.

The customs house at Cruces doubled as the home

of the *alcalde*, the mayor-cum-magistrate of the town. The mission would soon learn that *alcaldes* could be indispensable allies. The *alcalde* of Cruces entertained the travellers in his house and, after a day's rest, the two dozen members of the mission congregated beside mounds of baggage and instruments for the short cross-country trek to Panama. At eleven thirty on the 29th, a long column of laden mules plodded out of Cruces on the well-trampled track over the spine of upland separating the Atlantic from the Pacific. For La Condamine, it was a moment to recall the age of conquistadores: 'From the top of these mountains', he wrote, 'we saw for the first time the South Sea and the Bay of Panama, one of the most famous in the New World'. At six forty-five in the evening, they reached the sanctuary of Panama, where the president, Martinez de la Vega, received the scientists, 'particularly the foreigners [i.e. the French scientists] in the most cordial and endearing manner'.

After Portobelo, Panama was a breath of fresh air. Surrounded on three sides by the Pacific, the city was laid out on a peninsula, in a spacious, colonial grid, with a large open plaza and broad paved streets lined with single-storey timber houses roofed with tiles. The cathedral was built in stone and a defensive wall cut across the neck of the peninsula, adding to the sense of security. Martinez de la Vega was one of the most powerful men in Spain's New World colonies, a brigadier who had served a ten-year term as Cuba's governor before being posted in his late sixties to Panama as president of the *audiencia*. There was nothing lacking in the president's will to help the scientists on their way, but – yet again – the mission was stranded for want of transport. No available ships lay in port.

For weeks, they waited. January turned to February. Godin secured another tranche of pesos. Jorge Juan and Ulloa ordered some tents 'and other necessaries' for the forthcoming geodesic survey. When a merchant ship called *San Christoval* showed up, arrangements were made to pay for a passage south along the coast to Guayaquil. For both the French and Spanish members of the mission, the switch from sailing on the 'King's ships' under the command of qualified naval officers to sailing on a random merchant vessel was an uncomfortable adjustment. *San Christoval*'s captain, Juan-Manuel Morel, began unreliably. Departure dates were set and missed. Tensions inherent in a team too large to operate as a cohesive unit seethed to the surface each time a significant decision had to be made. One of the disagreements concerned the demand by La Condamine and Bouguer that the mission should pause on their passage to Guayaquil and undertake a reconnaissance on land close to the point where the equator crossed the coast. The two Academicians had become convinced by information gleaned in Panama that the hinterland beyond Cape San Francisco might be suitable terrain for the baseline and 200-mile chain of triangles that would be used to measure one degree of latitude. They had their eyes on an anchorage called Manta, just south of the equator. Godin refused to consider it. The surgeon, Seniergues, was so sick of the squabbling that he vented his frustration in a letter to Joseph de Jussieu's brothers in Paris, Antoine and Bernard de Jussieu. Sitting in a sweaty room in Panama, he wrote on 18 February:

> Godin disagrees with this and intends to go to Guayaquil and straight to Quito . . . La Condamine has already said in front of everybody that if no one else wants to stop there he'll stop on his own and if so *le Sieur* Bouguer will surely stay with him. *Le Sieur* Godin hasn't behaved well for some time – they scrap like dog and cat . . . it's impossible that they can finish this journey together.

Meanwhile, Joseph de Jussieu was writing to Antoine and Bernard with the list of misdemeanours committed by Godin, most scandalously the squandering of the king's money on a diamond and fancy clothing for a prostitute.

Godin was told that *San Christoval* would leave on the 19th, but the ship was still at anchor on the 20th. There was a palpable air of tension. This was the last leg of a succession of sea voyages that had been dragging on for more than nine months. Once they left Panama and the isthmus linking them to the Atlantic, the connections with Europe would be extremely tenuous. Their welfare would be entirely dependent upon their own survival skills and upon the hospitality of the colonial Spanish authorities.

Eventually, on 21 February 1736, the twenty-five members of the mission (a slave or servant had disappeared since Cartagena de Indias) trooped on to *San Christoval*. Next morning, they put to sea in weak and variable winds. It was a slow, uncomfortable departure as the ship was helmed south-south-westwards around the archipelago of low islands that lay waiting to trap any vessel trying to leave the Gulf of Panama. It wasn't until the 26th that they crept past Isla Iguana. Eventually, the ragged headland of Punta Mala – Bad Point – slipped from sight. Escaping the clutches of Panama's gulf convinced the scientists that the navigational skills of *San Christoval*'s captain were a hazard to all on board. The two Spanish naval lieutenants began keeping watch, taking their own star sights, logging the ship's speed, recording course changes, and ordering their own servants to take the ship's wheel when Morel's helmsman fell asleep.

It was on the voyage south from Panama that Godin managed to win back a few ounces of respect from some of his fellow travellers. His leadership skills were never in doubt: he was almost useless. But he had invested much time and energy in the scientific instruments that had been shipped from France. Among them was Hadley's octant, a revolutionary new device for determining latitude. Hand-held and taking the form of a 45-degree angle or one eighth of a circle (hence 'octant'), it used mirrors and scales to measure the height above the horizon of the sun and other celestial bodies. It could be used by day or at night and the two Spanish naval lieutenants had never seen anything like it: 'This ingenious gentleman,' wrote Ulloa, 'having been pitched upon for the voyage to America, undertook a journey to

London, purely to purchase several instruments.' The octant had been constructed by the instrument-maker John Hadley and it proved

> of the greatest use to us in finding the latitude during this passage; a point the more difficult and necessary, on account of several perplexing circumstances, the course being sometimes north, sometimes south, and the currents setting in the same direction. Assisted by this instrument, we were enabled to take the meridian altitude of the sun, whilst, from the density of the vapours which filled the atmosphere, the shadow could not be defined on the usual instruments.

After fifteen days at sea, they rounded Cape San Francisco and crossed the equator. All eyes followed the ragged green fringe of South America, searching for the headland known as Cabo Pasado: the 'last cape'. Beyond here, the coast receded in a broad bay and, in the crook of the next headland, they came to the small, sheltered anchorage of Manta. Bouguer and La Condamine were still keen to interrupt the voyage south to Guayaquil with a reconnaissance on shore, close to the equator. Aiding their cause was the ineptitude of *San Christoval*'s skipper, who had left Panama with inadequate provisions to reach Guayaquil. The ship needed a safe haven in order to take on fresh water and food.

On the afternoon of 9 March, *San Christoval* eased into the shelter of Manta Bay and dropped anchor in 11 fathoms of water. Next day, the mission went ashore and hiked past the ruins of Manta village – abandoned following successive pirate attacks – up the hillside to the village of Montechristi, a collection of stilted bamboo huts perched some 10 miles from the coast. It did not take long for the scientists to realize that the slopes above them were unsuitable for a geodesic survey. As Ulloa was to record, they 'soon found any geometrical operations to be impractical there, the whole country being extremely mountainous, and almost covered with prodigious trees'. Local people confirmed that the region was steep and forested. After a night on land, the mission retraced its steps to the coast and re-embarked on *San Christoval*, where Morel was supervising the loading of water and food. While the ship was at anchor, instruments were brought to the deck and the location of Manta was observed to be 56′ 5½″ south. They were less than one degree from the equator.

For Bouguer and La Condamine, the temptation was irresistible. From Manta Bay, *San Christoval* would take a week or two to reach Guayaquil, where they would have to wait for at least two months for the rains to ease and for the trail over the Andes to re-open to mule traffic. The debate that had been rumbling episodically since Panama now erupted. Godin was determined that the entire mission should stay with *San Christoval* and continue to Guayaquil. Bouguer and La Condamine wanted to disembark at Manta to conduct scientific observations on the equator and to extend the geodesic reconnaissance. 'It is already known,' wrote Bouguer, 'that we believed we might make some use of our time in this part of the coast, on which the heavy rains had already ceased to fall.' The scene was set for a mutiny.

Through the 12th, Jorge Juan and Ulloa found themselves acting as go-betweens, conveying messages between Bouguer and La Condamine up at Montechristi and Godin on board *San Christoval*. Godin scratched a testy two-page letter to the group on land, accusing his fellow Academicians of remaining in Manta without his consent and of 'having refused to obey orders'. He closed by announcing that he felt himself 'obliged to go as soon as possible to Guayaquil'.

On 13 March, *San Christoval* sailed out of Manta Bay, leaving Bouguer and La Condamine on the shore with their instruments, two slaves and one servant.

4

It was a chaotic debacle. The five castaways had been left without guides in an unmapped land. While the mood on shore was one of exploratory excitement, the atmosphere on board was tainted by Godin's humiliation. His flailing leadership had provoked a scientific insurrection. Before the geodesic survey had even started, the mission had fallen apart.

From Manta Bay, *San Christoval* rounded Cape San Lorenzo and set a course for Guayaquil. There was little in the way of useful science that Godin could achieve on the voyage, but a lunar eclipse was due on 26 March. If *San Christoval* could reach Guayaquil in ten days, and if the skies were clear on the night of the 26th, Godin would be able to determine the longitude of this key Spanish port. One of the instruments that he had sourced on his trip to London was a precision pendulum clock constructed by the well-known clockmaker George Graham. This large, sensitive device could be used to measure in seconds the duration of an eclipse, calibrated to local time. By comparing it with the time of the eclipse recorded in the Paris observatory, it would be possible to work out the precise longitude of Guayaquil. Every hour of difference would equate to 15 degrees of longitude. In the absence of Bouguer and La Condamine, the two Spanish lieutenants became Godin's willing partners. They, too, had an interest in determining the longitude of Spain's most important port in northern Peru.

Initially, *San Christoval* made good progress southward, coasting past the twin peaks of Isla de la Plata and then altering course to head south-south-east. Cape Blanco was passed on the 17th as the ship entered the Gulf of Guayaquil. At noon on the following day, Morel anchored half a league off the mouth of the Tumbes river, where *San Christoval* remained until the 20th, due – as Ulloa put it – to some 'particular affairs of the captain'. When Morel eventually weighed anchor, the current proved so strong that the ship was pulled back out to sea. The only way they could make progress towards Guayaquil was by sailing on the flood tide and anchoring on the ebb, then repeating the process. It wasn't until the 23rd that the ship drew level with Isla Puna and Morel was able to send for a pilot. The following day, he guided the ship into a small harbour near the island's northern tip. Only two days remained until the eclipse. Guayaquil lay 40 miles to the north, up an island-dotted estuary. Rather than miss the eclipse, Godin, Jorge Juan and Ulloa looked around the village beside Puna harbour for a structure that could be used as a makeshift observatory. But the walls of the houses were constructed from canes, which were insufficiently solid to support the sensitive instruments.

With time running out, the decision was taken to leave *San Christoval* in Puna harbour and to dash for Guayaquil in a rowing boat. Having loaded the instruments, Godin and the Spanish officers left Puna shortly before midnight on the 24th. Battling against an ebb tide, the river's current and darkness, the Puna oarsmen did their best, but it was not until five in the evening on the 25th that the exhausted crew bumped against Guayaquil's quay and lifted the instruments ashore. Summoning their remaining energy, the men managed to settle the pendulum in time for the eclipse but, as Ulloa morosely recorded, their 'diligence was entirely frustrated, the air being so filled with vapours, that nothing was to be seen'.

By evening the following day, *San Christoval* had caught up with them and lay at anchor before the city. Riverboats unloaded the mission's baggage and remaining instruments and Godin turned his mind to the next difficulty. From Guayaquil onwards, the mission would be operating in the jigsaw of districts – or *corregimientos* – used by the Viceroy in Lima to control the whole Viceroyalty of Peru. Each *corregimiento* was under the jurisdiction of a crown-appointed *corregidor*, who functioned as magistrate, judge and governor of his district. They were powerful, corruptible and their cooperation was essential if the mission was to reach Quito. With Jorge Juan and Ulloa in tow, Godin sought an audience with the *corregidor* of Guayaquil. They were received, wrote Ulloa, with 'great civility'. Orders were dispatched to all the *corregidores* through whose jurisdictions the mission hoped to pass on the way to Quito. The Guayaquil–Quito route was one of the Viceroyalty's main arteries, but it was also notorious for blockages. During the wet season, heavy rains and snowmelt turned fords into death traps, destroyed bridges and ripped away sections of trail. Yet again, the weight and bulk of the mission's baggage complicated the logistics. To reduce the number of difficult river crossings by mule, the mission would have to travel inland by river-boats to a place called Caracol, then transfer to mules for the mountain crossing to Guaranda, a small town in the province of Chimbo, which opened the way north to Quito. Outside the rainy season, the journey from the coast to Quito might take a month. But they would not be able to leave Guayaquil until the rivers were navigable and until mules had been sent by Guaranda's *corregidor* to Caracol.

Beside logistics, Godin had money worries. Paying for *San Christoval* to transport the mission and its baggage from Panama to Guayaquil had emptied the chest. And, looking forward, Godin needed to secure funding to cover the mission while it waited in Guayaquil for the waters to subside. Then he needed to cover the cost of transporting the mission from the coast to Quito, along 500 miles of river and trail. From the treasurer in Guayaquil, Godin obtained 2,100 pesos. Nearly three-quarters of the pesos had to be handed to Morel for the chartering of *San Christoval*.

Guayaquil was a less bad stopover than Portobelo, but it was no Paris or Seville. To Jorge Juan and Ulloa, this was the river-port founded by the great conquistador, Captain Francisco de Orellana, who went on to become the first European to make a recorded descent of the Amazon, a river named for a while on Spanish maps as Rio de Orellana. Guayaquil shared the Amazon's liquid demeanour. The entire place appeared to tilt on the west bank of the river like a decomposed shipwreck. Its houses, convents and churches, and even its three forts, were built of wood, a multitude of arks awaiting the seasonal floods. During the wet season, from January until June, torrid days were followed by drenching rain that turned the city's alluvium into a muddy slurry. Ulloa warned that the streets were 'not to be travelled over either on foot or horseback during the winter' and that the first rains turned them into 'one general slough' that had to be traversed on 'very large planks' which 'soon become slippery, and occasion frequent falls'. Worse than an involuntary mud-plunge was the venom of Guayaquil's wildlife. Snakes, scorpions and giant centipedes, observed Ulloa, 'find methods of getting into the houses, to the destruction of many of them'. It was, he added, 'absolutely necessary to examine carefully the beds, some of these animals having been known to find their way into them'. At night, rats entered the houses and ran up the walls and along the ceilings. Everybody slept beneath mosquito nets. It was impossible to keep a candle burning, except in a lantern, for more than three or four minutes before it was extinguished by 'numberless insects flying into its flame'.

Clouds and rain made astronomy almost impossible. 'The desire of succeeding rendered us very attentive to observe an immersion of the satellites of Jupiter, to make amends for our disappointment of the eclipse of the moon,' wrote Ulloa, 'but we were in this equally unfortunate; the density of the vapours which filled the atmosphere rendered our design abortive.' Whenever the rain ceased at night, the two Spanish lieutenants exposed themselves to the insects so that they could attempt to snatch observations through gaps in the cloud cover. The stings, remembered Ulloa, 'were attended with great tortures'. On more than one occasion, they were forced to abandon their labours, 'being unable to see or breathe'. Eventually, they fixed Guayaquil's latitude at 2° 11´ 21˝ south of the equator, but were unsuccessful in determining longitude 'by any accurate observations'.

Ulloa did manage some good geography in Guayaquil. With the same discipline that he had deployed in Cartagena de Indias, and in Portobelo and Panama, the young lieutenant devoted his days to urban exploration and compiled a detailed dossier on the human and physical geography of the city and its surroundings. His insights ranged from fashion to the biogeography of mangrove swamps and the architecture of giant balsa-wood rafts known as *balzas*: up to nine tree trunks lashed side by side, with a cabin of reeds and a mast formed from two poles of mangrove wood. Of the other members of the mission, Seniergues managed to make something of the long, torrid stopover in Guayaquil, treating one of the city's wealthy inhabitants for cataracts. La Condamine admiringly reported that the surgeon had 'made a considerable sum'.

Two months after disembarking *San Christoval*, the

mission was still in Guayaquil and the last pesos were pouring downstream like Andean silt.

At the beginning of May, word reached Guayaquil that mules provided by the *corregidor* of Guaranda were on the trail to Caracol. A scramble ensued as the mission's members evacuated lodgings and assembled their crates and bags on the muddy waterfront beside a large *chata* that would ferry them upstream. Swathed in mosquito cloths, Godin and his depleted mission left Guayaquil on 3 May and began following the lazy meanders of the river inland towards the Andes. The *chata*'s itching passengers had no idea that Pierre Bouguer had spent that day fighting through waterlogged rainforest in an attempt to catch them before they left Guayaquil.

The Manta mutiny left Bouguer and La Condamine ill equipped for an expedition on the coast of Peru. The day after the argument with Godin, *San Christoval* sailed south with virtually all of the mission's equipment. The following day, 14 May, Bouguer and La Condamine, with their two slaves and one servant, climbed back up to the village of Montechristi, where they were offered lodging in a large, stilted bamboo cabin described by Bouguer as the 'Casa Real' or 'King's House'. To reach its floor, the Academicians had to climb a ladder cut from two large bamboos 'in which they had contrived notches to receive one's feet'. The few items of equipment they had managed to bring ashore lay on the bamboo floor: 'I had only taken with me my instruments, a hunting suit and a hammock,' remembered La Condamine.

They had to take stock of their situation. This was a golden opportunity to do new science. Never before had two Academicians found themselves on the equator with instruments. Bouguer had managed to bring the pendulum ashore and a Réaumur thermometer. La Condamine had his compass and both of his quadrants. The larger one, with a three-foot radius, was a cumbersome beast that had once belonged to the great Chevalier de Louville, the first instrument-maker to fit a micrometer to a quadrant's telescope. La Condamine's smaller, more portable quadrant had a one-foot radius. Because Bouguer had been unable to bring his own quadrant ashore, La Condamine lent him the large Louville instrument. As important as the instruments was the French passport requesting local officials to offer 'all help, aid and favour' to the visiting scientists. La Condamine had translated the document into Spanish, and also brought from the ship a copy of the orders issued by the Spanish king. On the 15th, the bamboo cabin was visited by a group of local people headed by their *alcaldes* bearing their wands of authority. The bemused Academicians were presented with fruit and the good news that the lieutenant at Portoviejo – the regional town – had ordered that the French scientists were to be accorded 'the same attention as to himself'.

Bouguer and La Condamine wasted no time. About one third of a league from the village, high above Cape San Lorenzo, they selected the site for an observatory, which was roofed by their 'good friends the Indians with much facility'. The March equinox, when the turning

Earth would present the sun directly above the equator, was due to occur on the 21st, and Bouguer wanted to try a new method of recording its 'precise moment'. Unfortunately, the sun was obscured on the morning of the 22nd, so the observation was unsuccessful. Then an attempt to observe eclipses of Jupiter's moons was also obscured by cloud, but they did succeed where Godin had failed, in observing the lunar eclipse on the 26th and establishing the longitude of Cape San Lorenzo, an observation that enabled them to claim – mistakenly, as it transpired – that they had discovered the most western point of the continent of South America.

With a run of late nights completed at their makeshift observatory, the two Academicians travelled inland to meet the man who had facilitated their stay at Montechristi. Joseph de Olabe y Gomarra welcomed the visitors into his home in Portoviejo, offered to lend them money for their onward travels and provided much-needed local information. It may have been Olabe who provided the Academicians with the firearm, probably a musket, that La Condamine used in the rainforest several weeks later. In return, La Condamine administered some of his 'Jesuits' powder' to a man who had been suffering for the past year from recurring fevers. The two Frenchmen were amazed that their patient had never heard of this particular *fébrifuge*, or fever treatment, which was a native remedy from his own homeland. Jesuits' powder took its name from an old story about a missionary who had learned from Andean villagers about the powdered bark of the cinchona tree. The powder had been exported to Europe, where it was widely used in the moist lowlands of Spain and Italy, whose 'bad air' was thought to be the cause of deadly fevers known as *mal'aria*. La Condamine had brought some of the powder from France, knowing that he ran the risk of falling to fevers while travelling through the Caribbean and South America. The encounter at Portoviejo was the prelude to research into the cinchona tree that would occupy La Condamine – and others – for many years. Cinchona trees could be found on mountainsides in the district of Loja, some 300 miles south of Quito.

Relieved of the tensions created by sharing space with Godin, Bouguer also submitted to curiosities beyond the immediate demands of the mission. With the unfinished treatise on ship design in his baggage, he was fascinated by the timbers of the inland forests, from hard, black ebony to fragrant Guaiacum – a popular cure for syphilis – to a giant white tree whose timber was 'four or five times lighter than the lightest fir'. He had found the source of Ulloa's balsa. Nothing, wrote Bouguer, 'can be found more proper to make rafts'. Another tree, 'known under the name of Maria', also caught his eye. Noticeable for its tall, straight trunk and white bark, its timber was 'very flexible' without being 'excessively heavy'. Maria trees were valued on the coast by those who knew them as 'the only trees in Peru they can convert into masts for ships'. For sheer versatility, bamboo was one of Bouguer's favourites: a plant whose stems grew 'as thick as a man's leg' yet could be cut and trimmed for use as beams, joists and floorboards in houses held together by 'bark or rind, so that not a particle of iron enters into the composition or construction of the edifice'. For a man from the granite coast of France, the flexibility of Peruvian structures was disconcerting: 'Walk or move as gently as you can in these houses,' warned Bouguer. 'The whole edifice shakes.'

Convinced that the rest of the mission would be stuck down south in Guayaquil for many more weeks, Bouguer and La Condamine set off northward for the equator. From Portoviejo, it was a straight-line distance of 90 miles or so, but by horse, boot and dug-out, it was twice that distance. Local people provided the Academicians with horses and taught them how to 'profit of the flux and reflux of the tide' by riding along the hard, wet sand of the beaches instead of fighting forests and ravines inland. They passed through the Spanish settlement of Charapoto and, at the Bay of Caráquez, Bouguer's eye was drawn by the superb natural harbour and timber yards. Sometimes they rode and, at others, they were paddled along the coast in dugouts, or pirogues. Bouguer recalled that in more populated areas they managed to buy milk, eggs and poultry but otherwise 'subsisted upon rice, and what provisions we carried with us; the bananas and maize-cakes, which had no other fault than being exceedingly dry'. They skirted the great prow of Cape Pasado, and then the smaller headland of Punta Ballena: Whale Point. Then the land flattened where a meandering river – the Jama – squirmed through cut-off lagoons into the sea. They were only nine minutes of latitude south of the equator, a distance equivalent to a couple of days on horseback. But this was as far as Bouguer went. He had been looking for 'a commodious situation to observe the astronomical refractions near the horizon' and claimed that he 'at length found one at the mouth of the river Jama'. Here he stayed for fifteen days, watching the Earth turn as the evening sun dipped into the Pacific off Whale Point. No doubt he needed a break from La Condamine's restless enthusiasm, and he was also ill. Later, the data would enable him to compare refraction at sea level and at altitude.

While Bouguer gazed at the Pacific horizon, La Condamine pressed on, determined to reach the place where the equator crossed the coast. Some 15 miles beyond the River Jama, his quadrant told him that he had reached zero degrees of latitude, the imaginary ring that girdled Earth linking every point on the equatorial parallel. He was standing on a low, rounded hill that projected into the sea to form a slight promontory. In his journal, he identified the blunt headland as 'a point called Palmar'. The name came from the Spanish word for 'palm grove', but the brackish swamps around the hill made it a less than idyllic spot. The insects were ferocious and during the several nights La Condamine camped on the hill, persistent cloud made astronomy difficult. While he was there, he chose 'the most projecting rock' facing the Pacific and chipped a Latin inscription confirming that in 1736 astronomical observations of the 'PARIS SCI-ENTIAR ACADEMICA' had located the equator at 'PROMONTORIUM PALMAR'. Swatting the bugs, he recorded that the inscription was 'for the convenience of sailors', and that he 'should perhaps have added the advice not to stop there due to the persecution that people suffer day and night from mosquitos and various species of midges'.

Back together at the River Jama, the two Academicians had to make a tricky decision. A month and a half had passed since *San Christoval* had left them standing at Manta Bay. They had done some good science. They had reached the equator and placed it on a sketch map of the Pacific coast. Their astronomical observations and geographical fieldwork fulfilled a commitment to improve navigational knowledge along Peru's western seaboard. And they had confirmed that the coastal topography south of the equator was far too broken to set out a baseline and then conduct a geodesic survey over a distance of 200 miles. But they had run out of time. The rainy season was over and – as Bouguer observed – the roads 'were now beginning to be practicable'. If they were to catch up with Godin and contribute to the geodesic survey, they had to get to Quito. But they were separated from that city by at least 200 miles of tough travel through rainforest and mountains. The decision they took was commendably adventurous:

Being at this time at the mouth of the river Jama, which is nearly upon the same parallel with Quito, M. Condamine and myself agreed to separate and take different routes. M. Condamine followed the coast towards the north in search of the river of Emeralds, continuing to lay down a map of the country he crossed ... With respect to myself, retracing back my steps, I took a southern direction for Guayaquil, and penetrated the forests ...

And so, on 23 April 1736, the Geodesic Mission to the Equator became three separate expeditions, going in three different directions, out of contact with each other.

5

On the river above Guayaquil, Ulloa was being eaten alive. 'The tortures,' he wrote, 'from the mosquitoes were beyond imagination.' Outside the rainy season, when water volume was lower and flow slowed, a canoe could make Caracol in three days. Crammed on to the cumbersome, overloaded *chata*, fighting a swollen current, Godin's depleted team struggled upstream for eight days. There were several 'unfortunate accidents'. Ulloa did not specify their nature, but capsizes were not uncommon. The mosquitoes were so vicious that they bit through clothing. At night, sleep was virtually impossible. Itching, scratching and cursing, the men sought in vain to escape. In one dismal camp, they tried to shelter in an abandoned house, but the building seemed to attract every insect in the region. Some of the men left for the surrounding fields, where they faced both mosquitoes and snakes. Others tried burning the branches of trees, but the smoke choked them without deterring the swarms. Sunrise revealed swollen faces and bodies covered with 'painful tumours'. When they reached the small river-port of Caracol on 11 May, scientists and slaves alike were desperate for relief from torment.

Caracol occupied a slimy plot on the east bank and served as the transshipment point between *chata* and mule on the ancient route connecting Quito with the coast. For a couple of days, while they waited for the mules, the men stretched their legs, scratched their sores and prepared for the imminent trek. When the seventy mules from Guaranda eventually plodded into Caracol, a new problem emerged. There were insufficient beasts to carry both the mission's personnel and its multifarious barrels, bags and chests. The *chata* had been cramped but, most of the time, it had been a relatively safe platform for the mission's fragile instruments. Everything carried by a mule had to be lashed to the animal's back. The risk of damage was considerable, should the mule slip or allow its load to collide with a tree or a rock. In the chaos of muleteers and mud, the decision was taken to leave behind nearly one fifth of the baggage and equipment. It would have to follow when transport and conditions allowed.

Led by local guides, the mules and men began snaking eastward from Caracol. Savannah merged into woods of plantain and cacao and soon the mules were struggling through a saturated morass. Then they reached a ferocious little river called the Ojibar, cascading down from the Andes. After twelve months of little physical activity, the mission's personnel were not fit for an arduous trek. Muscles had wasted and feet had softened. For the two Spanish naval officers, the departure from tidal waters was an uncomfortable transition. They knew the sea and the rigidly controlled habitats of ships' decks. Within hours of leaving Caracol, they discovered how alarming and messy land travel could be. 'All the road from Caracol to the Ojibar,' observed Ulloa, 'is so deep and boggy that

the beasts at every step sunk almost up to their bellies.'
For two days, the convoy slithered and wallowed through
the rainforest, fording the river no less than thirteen times
(Ulloa was counting). The bridges were narrow, planked
structures with no side rails and a tendency to sway under
the weight of laden mules. At night, the local guides
hacked branches from the forest and constructed shelters
for the travellers to sleep beneath.

By the 16th, the trail was tilting upwards. They reached
a lofty waterfall that Ulloa found 'inconceivably beautiful'. They edged along alarming precipices. Men and
mules skidded and struck tree trunks and rocks. Loads
slipped and ropes had to be re-tensioned. Every hour
increased the tally of bruises. Delays were continual and,
for the Europeans, there was always the fear of being
accidentally isolated from the mule train and being left
alone in the wilderness.

On the 17th, they woke, feeling cold, at a place called
Tarigagua. Ahead rose the ascent into the mountain
region known to Spanish travellers as San Antonio. Later,
Ulloa reflected that this section of the journey gave the
Europeans more trouble and fatigue than anything they
had experienced in the past. The combination of rain,
mud and a route that ascended and descended repeatedly
along steep mountainsides disconcerted the usually fearless Spanish naval lieutenant. The trail was barely wide
enough for the mules to move and it was also beset with
deep holes that could leave a rider floundering waist deep
in slime. So much rain fell on the route that the guides
carried small trenching tools with which they cut

drainage gullies across the trail each day. Huge fallen trees forced mules to be unloaded then coaxed around each obstruction. According to Ulloa, there was much 'damage to goods'. Godin looked on with anxiety as the instrument boxes were passed from hand to hand. Loaded mules were urged down vertiginous slopes. At the lip of an abyss, each animal would halt and contemplate the drop, then assume a crouch and launch into a long, toboggan: 'All the rider has to do,' recorded Ulloa, is 'to keep himself fast in the saddle without checking his beast; for the least motion is sufficient to disorder the equilibrium of the mule, in which case they both unavoidably perish.' These gravity-assisted descents on mule-back were achieved with 'the swiftness of a meteor'.

Five days out of Caracol, the travellers stood atop the mountain barrier that had intimidated them for so long. It was a place known locally as 'Pacara', the 'gate' or 'narrow pass', a notch in the Andes that connected the coast with Quito. Nimbly, the guides led the exhausted, filthy convoy down the steep, slippery trail into the province of Chimbo and a new land. For the young lieutenant, Ulloa, their deliverance was accompanied by visions of distant Spain:

> After we had passed the mountains beyond Pacara, the whole country, within the reach of the eye, during a passage of two leagues, was a level and open plain, without trees or mountains, covered with fields of wheat, barley, maize, and other grain, whose verdure, different from that of the mountain, naturally gave us

great pleasure; our sight for nearly a year having been conversant only with the products of hot and moist countries, very foreign to these, which nearly resemble those of Europe, and excited in our minds the pleasing idea of our native soil.

As the mission descended to the plain, it was received like a returning sun god. The *corregidor* of Guaranda rode out to meet them, with the *alcalde* and a retinue of officials and servants. Chimbo's Dominican priest presented himself, with several of his order. Verguin, the methodical engineer from Toulon, described the scene in a letter home:

As thus we made our way, where all along the road the humbler people lined the banks on either side, and four young Indians, dressed in blue, with white belts and white kerchiefs about their heads, holding a long *bâton* in the hand with a kind of banner atop, encircled us letting out cries of joy after their manner.

As the cavalcade entered town, the mud-spattered travellers were amazed to hear bells ringing across the rooftops. Every house they passed seemed to resonate with the din of trumpets, tabors and pipes.

We were led to the Royal residence, where the *corregidor* resides, he having prepared our lodgings there, and where bunches of greenery had been put around the pillars that lined the length of a gallery. We had iced drinks and during supper there was an orchestra composed of harps and violins.

The *corregidor* explained that it was the custom of any town to welcome its visitors.

After two days of rest in Guaranda, the mission tackled the final obstacle on its long journey from Europe to the equator. The road east climbed from fertile fields, streams and stands of trees into a tight, steep valley of tufted grasses that eventually mounted at 13,000 feet a bleak plateau known in these parts as a *páramo*: a barren, icy upland above the sloping grasslands and below the snowline. These frigid deserts would become familiar to every member of the mission. In thin air, the bedraggled travellers plodded below the icy drapes of mighty Chimborazo, soaring almost 8,000 feet above the dusty road. On the morning of the 23rd, they emerged from a cramped hut to find the entire landscape coated with frost. They rode, wrapped in every available article of clothing, onwards along the flank of Chimborazo, then began to lose height until – at two in the afternoon – they reached 'a small, mean place' called Mocha, where they sheltered for the night.

As they rode stiffly out of Mocha on the 24th, the silvered peaks of Chimborazo and Tungurahua rose equidistant on each side of the trail. Ahead on the left, they could see the jagged crest of Carihuairazo, whose smoother, lower slopes suggested that it had once been a volcanic giant. No Europeans had ever climbed these peaks, which had for millennia played a commanding role in Andean mythology. Carihuairazo was said to have been destroyed by 'Father' Chimborazo while fighting for the favours of 'Mother' Tungurahua. The trail swung north for Quito and they joined the remains of the

Great Road: the 3,700-mile, handmade highway built to connect the Andean empire of the Incas. There were reminders that the place picked from a globe for being suitable for geodesic survey was telling them a far older story. Earlier, in a valley below Chimborazo, the riders had paused at the foundations of a structure they learned from the guides was 'an ancient palace of the Incas'. On the 28th, riding beneath the great cone of Cotopaxi, they passed more ancient ruins on the plain of Callo.

Each side of them, the parallel walls of two cloud-capped mountain ranges – cordilleras – stretched northward like an avenue towards Quito and the equator. The excitement was palpable. This great natural avenue between the cordilleras had always been the most likely location for the laying-out of the long chain of measured triangles. The peaks they were passing between would soon perform as lofty surveying stations connected by a lattice of sight-lines.

On the morning of the 29th, more than three weeks after leaving Guayaquil, the long column of dust-caked mules skirted the small, round-topped hill known by Spaniards after its bread-shaped profile as El Panecillo. And there before them were the pale walls of a small city. They had reached the haven they had sought for so long. In the annals of European exploration, Quito held a sombre fascination as the town founded in 1534 by Sebastián de Belalcázar after defeating the last great Inca leader, Rumiñahui, at the Battle of Mount Chimborazo. Belalcázar and his men were certain that Rumiñahui had hidden a vast hoard of gold, silver and platinum in the city. Since then, visiting Europeans were assumed to be treasure hunters.

The weary troupe rode into a place that bore few signs of wealth. Some of the buildings around the main square appeared to be falling down. Ranged along the four sides of the square were the cathedral, the episcopal palace, the town hall and the palace of the *audiencia*, the latter in a shocking state of disrepair. Even Ulloa, who strove when possible to dignify Spain's presence in South America, was obliged to describe Quito's main square as being 'rather disfigured than adorned by the palace' and its walls as so precarious that they 'continually threaten to demolish the parts now standing'. For the administrative hub of an *audiencia* that reached from the Pacific coast to the upper Amazon, Quito seemed somewhat dishevelled.

Godin and his depleted international mission were greeted by the president of the *audiencia*, Dionisio de Alsedo y Herrera, who showed them to a set of decrepit rooms within the palace. In Ulloa's version of events, they were entertained for three days 'with great splendour' and introduced to the bishop and canons, the provincial auditors, the head of the municipal government 'and all other persons of any distinction, who seemed to vie with each other in their civilities towards us'.

Godin would soon discover the limits of Alsedo's munificence. Of more immediate concern were the two Academicians he had left on the Pacific coast eleven weeks earlier. There was no word of their whereabouts.

6

The future of the mission lay in the hands of a psychopathic accountant. Dionisio de Alsedo y Herrera ran the *audiencia* of Quito with a keen eye on financial miscreants. Born to nobility in Madrid and expensively educated, Alsedo had served in the Court of Auditors in Lima, where he had developed a specialization in trade. He had a particular dislike for smugglers and for the British, who had captured him when the Spanish treasure fleet was intercepted off Cartagena de Indias. The flagship, *San Jose*, had exploded, sending 600 sailors and passengers to the Caribbean seabed. Her hold had been loaded with silver and gold. Alsedo had been taken to Jamaica then released on a prisoner exchange.

For Alsedo, Quito was less dangerous than a Spanish galleon, but it had not been an easy posting. For years, he had managed a territory whose capital had been built beside a live volcano, whose streets were split by earthquakes and whose society was bitterly divided between two castes of the colonial age: Spanish-born *chapetónes* and native *criollos*. Backed by Quito's *chapetónes*, Alsedo liked it to be known that he was tough on crime. His decrees were announced with town criers, trumpets and drums, and his executions and banishments were a civic performance. A couple of years before the mission showed up, the circulation of counterfeit coins was dramatically reduced when two forgers were hanged and burned in the main square. In 1733, Alsedo spent over 1,000 pesos improving Quito's prison with new iron shackles, chains, handcuffs and a whitewashed torture chamber. The president portrayed himself as the saviour of a royal asset, convincing the King of Spain that he had inherited a city rife with 'repeated crimes of perfidy, murders and robberies' and that his tenure had transformed Quito into 'a peaceful and tranquil republic'. When the mission arrived in June 1736, Alsedo's eight-year posting had a few months to run and the last thing he needed was complications.

Outwardly, the mission looked like a gaggle of disorganized incompetents. Two of the three French Academicians had disappeared. The rest of the mission had reached Quito in various states of ill health and dishevelment, minus much of their baggage, with insufficient funds to support themselves. And Alsedo had been sent evidence suggesting that the mission was involved in smuggling European goods into the Viceroyalty. One week before the mission arrived, word arrived in Quito that *Vautour* had been caught with an illegal cargo of imported textiles. Whether the mission was engaged in smuggling had yet to be conclusively established. In line with normal procedures, the baggage Godin had arrived with had been inspected, and the full complement of personnel had been carefully logged. But some of their baggage was still in Caracol.

Smuggling aside, Alsedo could see that the scientists

might be useful. A new survey of the city and an up-to-date map would add gloss to his presidency. Already, he had laid the ground by responding positively to Madrid's directive to 'provide every assistance' to the mission. When Godin had sent word from Guayaquil, asking for mules, it was Alsedo who had ordered their dispatch. When Godin arrived in Quito, Alsedo had laid on the three-day reception and accommodated the visitors in his palace. And he had paid to Godin the outstanding few hundred pesos of the 4,000-peso credit arranged in Cartagena de Indias. But for a man who habitually resorted to execution for financial irregularities, Alsedo was having to exhibit extreme restraint. As if the suggestion of smuggling was not bad enough, Godin had brandished the directive from King Philip V instructing the Viceroy of Peru and the presidents of all *audiencias* 'to assist the astronomers with all the amenities that they ask'. The Frenchman wanted Alsedo to arrange for the Treasury of Quito to loan the mission money for accommodation, transport and for incidentals, to be repaid by the government of France at some point in the future. For accommodation, Godin was demanding houses with gardens suitable for setting up astronomical instruments.

For the recently arrived members of the mission, there was much to learn. Health was a constant preoccupation and the notebooks of Jorge Juan and Ulloa filled with alarming lists of hazards. Venereal disease was so common in Quito that 'few persons are free from it'. More dangerous were the 'malignant spotted fevers and pleurisies' and a disease known as *peste* that caused convulsions, biting, delirium and the vomiting of blood. Of greatest concern was 'mal del valle, or vicho', which often seized its victims two or three days after a fever. The mission's doctor, Jussieu, was convinced that vicho was 'gangrene in the rectum' and prescribed a treatment that 'must be attended with no small pain, as a pessary, composed of gun-powder, Guinea-pepper, and a lemon peeled, is insinuated into the anus and changed two or three times a day, till the patient is judged to be out of danger'.

Among the other hazards the mission faced were inflated costs and corruption. Everything from drinking glasses to imported cottons, wools and silks changed hands in Quito for higher prices than in France. Iron, which would be essential for the repair and construction of instruments, cost six reals a pound. Brass was virtually unavailable. Vexing, too, was the cost and complications involved in hiring the mules that would be essential for carrying the mission's personnel, instruments, tents and baggage during the geodesic survey. Mule procurement was in the firm grip of the *corregidores*, who ran their district transport systems through exploitation. Every year, the *corregidor* visited the towns and haciendas within his *corregimiento*, demanding tribute from everybody aged between eighteen and fifty-five (with certain exceptions for ill health or disability). Jorge Juan and Ulloa began keeping a systematic record of corruption in the Church and state.

Another tool of exploitation was the *repartimiento* system, in which communities were forced to acquire essentials such as clothing and food from the *corregidores*, who also set the price. Mules that might have cost the *corregidor* 14 to 16 pesos each were sold for 40 pesos. The double-bind was that muleteers had to seek permission before hiring out their services, and to pay their *corregidor* a cut of their takings. Transport was going to be a recurring problem for the mission.

On 4 June, six days after Godin had arrived in Quito from the south, another ragged French astronomer staggered into town, from the north. It was La Condamine.

When Bouguer and La Condamine had parted company on the coast, they swapped quadrants. La Condamine took the heavy, cumbersome three-foot quadrant, leaving Bouguer with the more portable one-foot instrument. Neither of them foresaw the scale of difficulties they faced in reaching Quito. Bouguer had the less dangerous, though longer, route, part of which he had already travelled. But he was unwell.

Accompanied by one slave, Bouguer retraced his steps back down the Pacific shoreline, then turned inland to Portoviejo and the forested uplands, with every expectation of reaching Guayaquil before Godin departed for Quito. Already weakened by illness, Bouguer soon ran into trouble. The forest trails were flooded and frequently knee-deep in water. Precariously clutching the saddles of their lurching mules, they pushed on through country still saturated by the rainy season. Bouguer – who was more used to teaching marine science than performing equine stunts – remembered a 'continual morass or slough' in which 'violent efforts by the mules to extricate themselves' threatened repeatedly to have their riders 'dashed against a tree'. The two men struggled into Guayaquil to find that Godin had left several days earlier. Without pausing for a rest in the city, Bouguer 'quitted it the same day', boarding a canoe for the river voyage upstream in a desperate attempt to catch up with the mission. When he reached Caracol on 19 May, he was told that Godin had left 'about three days before'. And he had taken all the available mules. Left behind in Caracol was a mound of mission baggage. Bouguer was spent. He was exhausted and ill and there was no transport available to help him chase after Godin.

La Condamine was not having an easy time either. The decision to separate at the River Jama bore his fingerprints. Travelling together, they could have pooled their energies and perhaps caught up with Godin and the mission before their departure from Guayaquil for Quito. But La Condamine was a rogue player. His fearless curiosity would always place him at the centrifugal periphery, poised to spin away from the centre into the fascinating void. Whenever he did this, the centre was destabilized. La Condamine was an agent of disequilibrium. The tensions he created kept everyone on edge and, on occasion, sharpened a keenness to succeed. He was any leader's bad dream. He had taken the three-foot quadrant because he wanted to continue filling in the blanks on the map of northern Peru. He was also curious about the Esmeraldas, the 'River of Emeralds'.

Predictably, La Condamine's urge to be an explorer led to difficulties. His two companions – his servant and a slave – were undoubted assets, but they were encumbered by baggage and the large quadrant. La Condamine tried to hire guides, but no one was prepared to leave home and accompany the Frenchman into the forest. So the three men began their explorative journey to Quito by sea-going dug-out canoe, paddled and piloted by local boatmen. Beaching at night and paddling by day, the tiny craft bobbed northward, with interruptions while La Condamine landed with the quadrant to fix the latitude of prominent coastal seamarks. The most important of these was Cape San Francisco, the promontory they had observed from the deck of *San Christoval* so many weeks earlier. Beyond the cape, they landed to fix the latitude of the river estuary of Atacames. About 10 miles further along the coast, the men in the canoe saw the seawater turn from translucent blue to muddy brown and, presently, a gap in the green shoreline appeared, where the Esmeraldas poured into the Pacific. La Condamine recorded that they had 'rubbed shoulders with the land in a *pirogue* for more than fifty leagues', a distance roughly equivalent to canoeing between Calais and Dover four times.

Entering the estuary, their troubles began. A huge amount of water was backed up in the Andes, turning the Esmeraldas into a *'rivière très-rapide'*. Making headway against the current was an exhausting struggle. The quest for precious stones was abandoned. On his map, La Condamine marked the approximate location as 'the small hill of the Emeralds'. Snake-like meanders drew them into the ancestral lands of the Nigua, tracts of rainforest and clearing that extended all the way up the Esmeraldas to the river's rising in the Andes on a volcano La Condamine came to call 'the Vesuvius of Quito'. Pichincha rose like a beacon from the western edge of the city. If they could reach Pichincha, they could find Quito.

Close to a valley known to early colonists as Puerto Quito, La Condamine and his two companions swapped the dug-out for horses and took to forest trails. The likelihood of onward progress now lay in the hands of local Nigua, without whose help they could neither carry their baggage and quadrant nor find the trails that would lead to Pichincha. La Condamine had a compass, and knew that Quito lay south-east of their presumed location on the Esmeraldas, but no human being could trek in a straight line on a compass bearing across rivers and ravines. The key to their survival lay in the knowledge of local foot trails that avoided the worst obstacles and in local porters to help them carry the baggage and instruments.

Initially, they were lucky in finding guides who could lead them along the rainforest trails and carry the instruments and baggage. Machetes were needed to clear the way. Intent on mapping, observing and collecting samples, La Condamine was both navigator and scientist. Interesting seeds were packed in his bags and his notebooks filled with drawings of rainforest flora. 'I walked with a compass and thermometer in my hand,' he recalled, 'more often on foot than on horseback.' Every afternoon, the rains rattled the rainforest canopy. The two hired guides 'had much trouble' with the quadrant. Before long, they abandoned their portering role and disappeared into the forest. With limited experience of rainforest survival, the trio made progress slowly. For four days, they pressed on, dependent upon La Condamine's compass and his reading of the terrain. Overloaded and losing flagons of sweat every day, they subsisted by shooting wildlife. When the powder ran out, they survived on bananas and wild fruits plucked from the forest. La Condamine became feverish. As they gained height, rivers became ravines. They followed a trail 'bordered by precipices hollowed out by torrents of melted snow, which fell in a great noise'. La Condamine was astonished by his first slender, swaying bridge of liana, sagging across a ravine 'like a fisherman's net'. In the coming years, he would become familiar with Andean bridges but, on first encounter, they seemed alarmingly unsafe. They reached a village and were assured that the path did indeed lead to Quito. By this stage, the money had run out and La Condamine had to accept that they were in trouble. He was forced to leave the quadrant at a hamlet, in return for a promise that mules would be sent to meet them at a village called Nono. When the three emaciated men staggered into Nono, they knew they were safe.

Nono lay one day's riding north of Quito, in a deep, lush valley of hummingbirds and pastures. The ragged, reeking member of the French Academy of Sciences managed to obtain some clothes on credit from a Franciscan monk. Revitalized, La Condamine set his sights on the beacon he had been thinking about for so many weeks. Nono's long, deep valley carried his gaze southward to a rising, grassy spine, angled to the sky. He would be first in the mission to climb Pichincha.

Up on the volcano, in cool, clean air above the rainforest, he recovered his sense of wonder:

> As far as my eye could see were cultivated lands, diverse plains and meadows, green hillsides, villages and hamlets surrounded by hedges and gardens: the city of Quito, far away, was at the end of this beautiful prospect. I felt as if I had been transported to the most beautiful provinces of France . . .

On 4 June, La Condamine entered Quito to find that Godin, with the majority of the mission, had arrived six days earlier. But Bouguer had not been seen by anyone since he parted from La Condamine on the Pacific coast in late April.

Stuck in sweltering Caracol, the professor of hydrography was in a bad way. While recovering and waiting for mules, Bouguer looked through the dumped mission baggage for his possessions. He also selected two of La Condamine's chests, which he hoped to take with him to Quito. A week later, he left the river on the long and grinding trail into the mountains. Travelling in a smaller group than the mission that had preceded him, Bouguer made good time but was also alarmed by the many crossings of the roaring Ojibar River and by the 'infinity of different precipices' he was forced to negotiate as the trail ascended. For the first few days the rain was constant and they were unable to light a fire. Food was restricted to 'bad cheese, and biscuit made partly of maize'. After a rest in Guaranda, he began the gruelling climb past Chimborazo and the *páramo* that 'beheld nothing but snow or hoar frost'. Then he descended into the land that had enchanted Ulloa a fortnight earlier:

How I was surprised by the novelty of the view! I imagined myself, after having been successively exposed to the ardour of the torrid zone and the horrors of the cold, transported all at once, as it were, into the temperate climate of France, and into a country, as embellished here, in the most engaging season.

Bouguer's geometrical imagination was on fire:

The houses, no longer constructed with bamboo, as are those lower down, but built of solid materials, some of stone, but for the most part of large bricks dried under shade. Every village is ornamented with a square, one of the sides of which is partly taken up by the church; in no region of the world have they failed to set this place, which is a parallelogram, to the east, from which streets divide in straight lines, open to the distant country; even the fields are frequently intersected thus at right angles, which give to them the form of a garden.

He arrived in Quito six days after La Condamine, on 10 June.

After an extraordinary succession of splits and

misadventures, all twelve of the core scientific team were in the same city. Godin had managed to secure from Alsedo a commitment to request funds from Lima. Alsedo had also arranged for the mission to use a pair of houses a few blocks north of the main square in the parish of Santa Barbara. Both houses had gardens suitable for the placement of astronomical instruments. Only La Condamine was missing.

On his arrival in Quito from his adventures in the rainforest, La Condamine discovered that most of his luggage, including his bed, had been left behind in Caracol. All he had were the two chests that Bouguer had kindly picked from the mission's dump on the other side of the Andes. Worse, his precious three-foot quadrant was stuck in a forest hamlet behind Pichincha. With no clothes and no instruments, he was literally and scientifically naked. Verguin lent him 50 pesos to pay for his quadrant to be rescued and taken to Nono, to await collection. Having ruled himself to be 'out of condition to appear in public with decency', La Condamine declined the opportunity of living with the other members of the mission and instead settled into the Jesuit seminary, just off the main square, adjacent to the president's palace. Here, he was afforded the privacy, time and space to record his explorations of the Peruvian coast and the Esmeraldas route to Quito. Very soon, he would discover that equilibrium in Quito was illusory.

7

The journeys to Quito had forged a unity of purpose. At last, they were ready to create the virtual chain of triangles from which they could compute the length of one degree of latitude at the equator. They would need to make maps of the great valley into which the first triangles would fit and then use the maps to identify inter-visible high points that would form the corners of each triangle. And they would need to establish a baseline by measuring the length of one side of one of the first triangles. That baseline would be the start line: if you knew the length of one side of a triangle, and knew two of its internal angles, you could compute the remaining angles and sides using trigonometric formulae. Once they'd measured the baseline, all the measurements would be of angles between the sides of triangles. The length of the entire chain could then be derived from the angles. The maths was simple. The measuring was monstrous. Large, visible 'signals' would need to be placed on the high points, many of them mountain summits. Only when two signals were simultaneously visible from a third signal – the observational 'station' – could the angle be measured, using a quadrant. Most stations would need an adjacent camp that the surveyors could use for shelter and provisions while they waited for clear weather.

All were now agreed that they had found the geodesic stage for their great project. The equator lay less than a day's ride north of Quito. South from Quito stretched the long, open-floored corridor, walled by peaks. The corridor extended southwards from Quito to Rio-bamba, a distance of around one hundred miles and equivalent to more than one degree of latitude. But in order to increase the accuracy of the final figure, Godin and Bouguer wanted to measure three degrees of lati-tude, which would mean extending the chain of triangles beyond Riobamba, into a region of crowded mountains. The distance to be surveyed would stretch for over 200 miles, from a point north of Quito to a southern termi-nus somewhere beyond the town of Cuenca.

The first act in this geodesic epic was the search for a site on the equator suitable for the all-important base-line: the line on the ground that would anchor the northern end of the chain of triangles. Zero degrees of latitude stretched like a taut rope from La Condamine's inscribed stone on the equator at 'Promontorium Pal-mar', through the rainforests and cordillera and across the volcanoes and valleys 20 miles north of Quito. The day after Bouguer arrived in Quito, two members of the mission departed on a baseline reconnaissance. In com-mand was the most experienced surveyor on the team: the cartographer-engineer Jean-Joseph Verguin. Acting as his assistant was young, ever-eager Jacques Couplet-Viguier, excited to be tasked with an opening role in the survey. They had been instructed to locate a level plain with a north–south axis at least 4,000 *toises* – around 5 miles – in length. It was a tall order amid the mountains of the Andes.

Verguin, Couplet and their guides rode north into the deep canyon of the Guayllabamba River, then climbed up towards the shattered summit of a volcano called Mojanda. Here, at an altitude of nearly 10,000 feet, stood the ruins of Cochasquí, a complex of fifteen pyramids built from volcanic stone long before the Inca civilization colonized the Andes. For as long as anyone could remember, Cochasquí had been embedded in the oral histories. It was the defeat of Quilago, the warrior queen of Cochasquí, that had allowed the invading Inca to take the surrounding valleys. Cochasquí's overgrown temples stood on the equator and may have been used in part as astronomical observatories.

Below the temple site stretched a sloping plateau occupied by the village of Malchinguí. It was the largest area of unbroken ground that Verguin and Couplet had seen since leaving Quito. Riding the hot, south-facing slope, the two men found that the plateau lost some 1,300 feet in altitude as it tilted downward towards Quito. It was framed to the north and west by the steep slopes of the volcano, and to the south and east by the abrupt edges of bluffs. But the plateau at Malchinguí stretched for only 3,000 *toises*, less than 4 miles. It was too short for a baseline.

From Malchinguí, they rode east along the equator, across the gullies and ravines cutting down from the volcano above their left shoulders. Then the land levelled. The plain of Cayambe had been settled by native peoples long before the Incas or the Spanish came to the land of the volcanoes. Initially, the plain looked promising for geodesy. A small town slumbered in the centre of a patchwork of green fields. But snowmelt and rain from the surrounding mountains had incised the plain with two deep rivers and numerous *quebradas* – dry ravines hacked by flash floods. The level space at Cayambe was no more extensive than that at Malchinguí. The two men turned and began riding back to Quito, bearing news that there was nowhere on the equator suitable for a baseline.

In Quito, Godin had problems.

The reappearance of Bouguer had triggered a fresh axis of initiative within the mission. With their shared backgrounds in hydrography and the sea, Bouguer and the two Spanish naval lieutenants rekindled the mutual interests that had developed during the voyages on board *Vautour* and *San Christoval*. Working in the garden of the Santa Barbara observatory, the mathematically minded trio established new, more accurate figures for the latitude and longitude of Quito.

La Condamine, meanwhile, was writing himself into the Academy's records. While waiting for his quadrant to be brought to Nono, he was hunched over a table in the seminary compiling a map of the Peruvian coast and another of the route he had taken up the Esmeraldas and then through the mountains to Quito. He attached to the maps a detailed account of the many observations he had taken on the coast with Bouguer, then produced two copies of everything to send to France, one to the Academy and the other to Maurepas.

Among the episodes that La Condamine was keen to commit to paper was his discovery of a *résine élastique*, a pale, resinous sap that was being tapped from trees in the rainforest. The sap was collected in banana leaves and then allowed to dry until it had attained a malleable consistency. Moulded into torches, the sap would remain aflame, even in the rain. And when formed into unbreakable gourds, it could be used for carrying water or juices. La Condamine was fascinated by this versatile, natural compound. Among its unique properties, a ball of the sap bounced when dropped on to a hard surface. Excitedly, he wrote to his Academy friend Charles du Fay with a detailed description of the sap and a package of samples. Much later, La Condamine would be credited with having made the first European discovery of rubber.

For a while, La Condamine ran his own show from the seminary. He made himself at home by erecting a nine-foot rod in the seminary courtyard, with a meridian on the ground that could be used to determine noon on sunny days. The gnomon in the Quito courtyard joined the inscription on the rock at Palmar as minor monuments on La Condamine's erratic course towards geodesic fulfilment. Unencumbered by the constraints of team etiquette, the Academician received a succession of visitors, among them Ramón Joaquín Maldonado y Sotomayor, a young, influential member of Quito's elite who was eager to promote the idea of a direct road link between Quito and the coast by way of the Esmeraldas. For Maldonado, the sudden appearance of French astronomers in Quito was an opportunity too great to miss. While Verguin and

Couplet were searching for a baseline up north on the equator, Maldonado took La Condamine on a private excursion to Nono, so that the Academician could collect his quadrant and see the route of the proposed road, which dived between volcanoes towards the Esmeraldas and the coast. Before leaving Nono, La Condamine used the quadrant to fix the village's location, finding that they were just one minute of latitude north of the equator. When he got back to Quito, La Condamine discovered that he was not the president's favourite Frenchman.

Ever since the team had reunited in Quito, Charles-Marie de La Condamine had been making a mild nuisance of himself. By declining the mission lodgings in the parish of Santa Barbara and instead settling into the Jesuit seminary, he had irritated Alsedo, who accused him of 'being disunited in carrying out your commission to measure the exact figure of the Earth'. Alsedo was also angered by La Condamine's unauthorized journey up the Esmeraldas to Quito and had reported the misdemeanour to Villagarcía, the Viceroy of Peru. La Condamine was also undermining Godin's leadership. In a few short weeks of freelancing on the coast with Bouguer, the pair had achieved more than the leader of the mission. And now La Condamine had found an influential ally in Maldonado. La Condamine was an ungovernable force, but he also knew who to please. The wayward Academician apologized to Alsedo for the slights to his presidential authority and moved into the observatory buildings in Santa Barbara.

The failure to find a suitable plain for the baseline was

another disappointment for Godin. Without a baseline, he could not begin to demonstrate to Alsedo that the mission was worthy of support. And by the end of June, there were insufficient funds for the mission to continue work. Bills of exchange Godin had expected from France had not arrived. All he could do was wait upon a reply from the Viceroy in Lima. The eventual, curt response was dated 21 July. In it, Villagarcía informed Alsedo that any payments made to the mission would have to come directly from the crown attorney in Quito. But Alsedo had already made it clear to Godin that he could not expect any pesos from his *audiencia*.

Effectively, the mission was broke. Fifteen months after leaving France, the scientists had managed to assemble the complete team at the equator, with all the necessary instruments. All they lacked was pesos.

Charles-Marie de La Condamine had been waiting for this moment. As the mission foundered, he announced that he possessed letters of credit from the Castanier bank in Paris. Their worth was roughly equivalent to 20,000 pesos, dwarfing by far the 4,000-peso credit that Godin had secured in Cartagena de Indias. Through Castanier, La Condamine could provide the mission with sufficient funds to conduct the survey. But there was a snag. To secure the pesos, La Condamine would have to travel in person to Lima, a round trip from Quito of at least three months. A plan was devised. If the mission could raise provisional funds by selling certain possessions to Quito's merchants, the measuring of the

1. In 1735, scientists sailed from Europe to a land they could only imagine, where the equator cut through rainforest, ravine and snowy volcano. 'Heart of the Andes' is an imaginary view of what is now Ecuador, by the American landscape painter Frederick Edwin Church.

2. Louis Godin, the expedition's nominated leader, clutching the world he lost. Godin returned from South America long after his companions, discredited and forgotten.

3. Pierre Bouguer, bilious in lilac, nine years after he returned to Europe to be captured in pastels by Jean Baptiste Perronneau. He had not wanted to join the expedition and disliked sea voyages, yet became the mission's leader.

4. Charles-Marie de La Condamine looking mischievous in a pastel portrait of 1753 by Maurice Quentin de la Tour, the celebrated painter whose long list of subjects included La Condamine's friend, Voltaire, and King Louis XV.

5. Jorge Juan y Santacilia, the senior of the two Spanish lieutenants on the mission, seen here as a decorated national hero. He returned from South America to become a spy in England and a trouble-shooter in Morocco and Spain.

6. Antonio de Ulloa y de la Torre-Guiral was nineteen when he sailed from Cádiz for South America. A compulsive writer and geographer, his published accounts of the expedition included a damning report on humanitarian crimes in the Spanish colonies.

LIBRO TERCERO.

Viages desde el Puerto del *Callao* à *Europa*; con noticias de las Navegaciones desde la *Concepcion de Chile* hasta la Isla de *Fernando de Noroña*, *Cabo Bretòn*, *Terranova*, y *Portsmouth* en *Inglaterra*, y desde el mismo Puerto del *Mar del Sùr* al de el *Guarico* en la Isla de *Santo Domingo*, y de este al de *Brest* en *Francia*.

7. In the globalized world of the 18th century, information, goods and people moved along a web of shipping lanes exposed to inaccurate charts, storms, pirates and marauding enemy fleets. The mission's journey to the equator took nearly one year. The illustration is taken from *Historical Relation of the Voyage to South America*, the book by the mission's two Spanish naval lieutenants.

8. First landfall for the French contingent was the island of Martinique in the Caribbean, where La Condamine suffered a violent fever known as 'Black Vomit'.

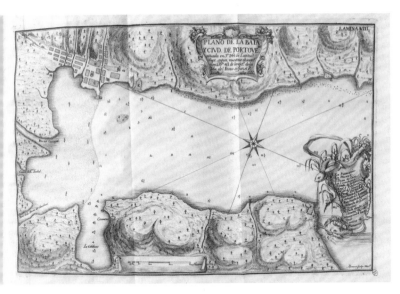

9. Portobelo hell: for three weeks, the mission was forced to wait for riverboats in a place so rife with insects and illness that it was known as 'the graveyard of Spaniards'. Jorge Juan and Ulloa managed to draw this detailed chart of the anchorage and town.

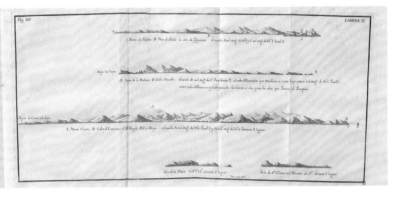

10. These coast profiles were drawn by Jorge Juan and Ulloa during the sea-passage from Panama to Guayaquil. On the upper profile is Bad Point (B) and the Isle of Iguanas (C). On the second profile, 'A' is 'Whale Point', where Bouguer waited while La Condamine paddled north in search of the equator. 'B' marks the rocks of Last Point. On the third profile, 'A' is the peak of 'Monte Christo', above Manta Bay, where Godin marooned Bouguer and La Condamine. The lowest profile shows the two islands that must be passed by ships sailing from Manta to Guayaquil.

11. La Condamine the Explorer at 'Punta Palmar', where he decided the equator cut the Pacific coast. Behind the rock he is inscribing are his quadrant, the pendulum clock and a map in the making. The hammock, shelter and canoe remind his readers that science requires sacrifice.

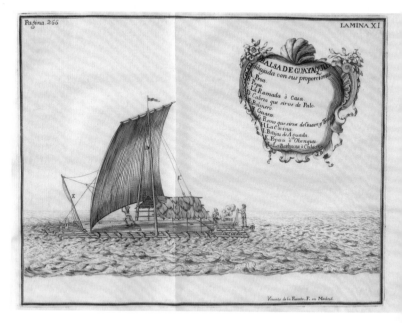

12. Balsa rafts fascinated the mission's two naval lieutenants. In their diagram, the raft is held on course by two retractable keels (F) and has a sail supported on an A-shaped mast of mangrove wood (D). The open fire on the stern serves as the kitchen (H).

Demostracion de la Montaña de S.ᵗᵒ Antonio, con la de los precipicios y riesgo de su Camino.   Mexico sculp.

13. The trail over the Andes teetered above roaring gorges. On steeper sections, mules 'tobogganed' downwards, while guides clung to trees ready to snatch riders to safety.

14. Rivers crossings needed nerves of Toledo steel. Gorges had to be crossed on swaying bridges of rope, or in slings hauled from the far bank.

A. Guabas ó Pacaes= B. Aguacate= C. Chrimoyo= D. Granadilla= E. Frutilla ó frsa de Quito=
F. Llama= G. Muca muca=H. Danta ó ozan Uestia= I. Quinual= K. Achupalla=L. Palo de luz
M. Puc-huchu=

15. Andean flora and fauna, from the book by Jorge Juan and Ulloa. The avocado tree (B) had 'the choicest fruits' while the cherimoya (C) was 'most delicious'. Wildlife included the llama (F) and 'muca-muca', or opossum (G). Not shown are the snakes, scorpions, mosquitoes and jaguars.

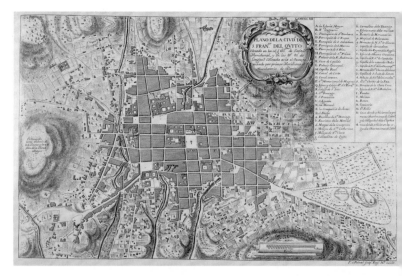

16. Morainville's map of Quito, with north-west at the top, where the suburbs meet the volcano Pichincha. The main plaza, where Jorge Juan killed the president's secretary, is shown at the centre of the colonial grid of streets. The mission's Santa Barbara observatory was located three blocks north of the plaza. The hill of Panecillo, where cannons were fired for the speed-of-sound experiments, is on the left of the map.

baseline could be undertaken during the fine weather before the onset of the rainy season in November. La Condamine could then travel to Lima during the wet months. For La Condamine, it was a coup: he could be elevated to saviour of the mision, he could undertake a journey of exploration through the land of the Incas, and then he could overwinter in Lima, where he could rub shoulders with Villagarcía, the Viceroy of Peru.

The financial rescue plan moved with remarkable fluidity, as if La Condamine had been prepared in advance. A document stating the terms and conditions related to the bank loan was produced and signed by La Condamine, Godin and Bouguer. And from La Condamine's quarters emerged an astonishing variety of items 'surplus to requirements': needles, bullets, several lace shirts, an expensive musket and 'several pieces of furniture'. One witness claimed to have bought directly from La Condamine a 'brilliant ring and a cross of St Lazarus enriched with diamonds'. Later, La Condamine explained that 'both masters and servants' on the mission 'had sold, in order to meet our present needs, the things which we could do without'. La Condamine's friends at the Jesuit seminary hosted the sell-off, which produced sufficient funds to cover the next few months of mission activities.

At the beginning of September, they returned to the search for a baseline. Godin, it seems, was arguing for the plain at Cayambe, despite its topographic problems. Not for the first time, La Condamine had another idea. But the clock was ticking. If they could agree on a baseline, they would have an opportunity provided by a lunar eclipse on 19 September to fix the longitude of the line's two ends. Mules were hired. On 10 September, Godin, Bouguer and the two Spanish lieutenants rode out of Quito, taking the circuitous trail north then east towards Cayambe. La Condamine chose a different route.

Two days after leaving Quito, Godin's party reached Hacienda Guachalá, at the southern tip of the Cayambe plain, where they unloaded the mules and became acquainted with the estate's cooperative owner, Antonio de Ormaza. Some of his buildings went back to the first generation of Spanish settlers in the 1500s. For Godin and Bouguer, it was an opportunity to stand for the first time on the equator, which crossed Ormaza's land just north of the hacienda. But it was clear that Verguin and Couplet had been correct: Cayambe's plain was small and the two rivers were geodesic gulfs. Godin and Bouguer began searching for level terrain that could accommodate a straight line of at least 4,000 *toises*: 5 miles.

La Condamine was still one day's ride to the south. When he left Quito, he had taken an easterly route, descending from the city to the valley floor then crossing the grain of the country, where the road scrambled across a number of flat-topped ribs tilting down from the heights of the eastern cordillera. It is possible that La Condamine knew precisely where he was going, either because he had seen the plain of Yaruquí from the heights of Pichincha, or because he had been told about it by his new friend, Ramón Joaquín Maldonado, whose brother José Antonio Maldonado was the parish priest at El Quinche, an old Inca town close to Yaruquí. That day, La Condamine found himself gazing from the El Quinche road down a long, thin, gently sloping plain rimmed by steep drops along its sides and far end. It was on roughly the same latitude as Quito but lower in altitude, and angled slightly to the north-west. With the exception of a sinuous *quebrada* on the eastern side of the plateau, Yaruquí's plain appeared unobstructed. And it was far more than 5 miles long.

Without wasting a moment, La Condamine hurried on to Cayambe, where he found that Bouguer 'had just recognised one of the extremities of the planned base'. The two men compared notes. Bouguer was unhappy. He 'found the ground very uneven' for the baseline and needed little persuading that Cayambe was a lost cause. Even Godin wanted to see Yaruquí, La Condamine writing that the expedition leader 'had also heard of this plain'. Godin, Bouguer and La Condamine rode south and spent two days thoroughly investigating the plain at Yaruquí.

La Condamine was right. The surface of the plain was not smooth, and lost some 800 feet in altitude as it dipped northward, but the slope was acceptable and if a clear path could be scraped in the vegetation and through several walls it would be possible to create a baseline about 7 miles in length. With no time to waste before the lunar eclipse, Bouguer, La Condamine, Jorge Juan and various unidentified assistants marked the line with poles and identified its two ends, Caraburo in the north and

Oyambaro in the south. By the 19th, they were ready to observe the lunar eclipse and fix the longitude of their baseline. That night, the moon was almost overhead. But as the sky darkened, one of their companions lay dying in Cayambe.

8

The eighteen-year-old had been feeling unwell before the mission rode out of Quito. But nothing was going to keep him from the inaugural measurements of the geodesic survey. He was young and tough and everybody expected him to bounce back to good health. But a week after leaving the city, he was on a bed, slippery with sweats, and the people who might have helped were elsewhere: Jussieu, the mission's doctor, was back in Quito and La Condamine – whose Jesuit's powder had cured a man in Portoviejo – was miles away at Yaruquí, focussed on an eclipse.

At the hacienda near Cayambe, Antonio de Ulloa watched his friend slip into unconsciousness:

> In this place we lost M. Couplet... He was indeed slightly indisposed when we set out from Quito, but, being of a strong constitution, his zeal for the expedition would not permit him to be absent at our first experiment. On his arrival, however, his distemper rose to such a height, that he had only two days to prepare for his passage to eternity; but we had the satisfaction to see he performed his part with exemplary devotion. This sudden death of a person in the flower of his age, was the more alarming, as none of us could discover the nature of his disease.

Jacques Couplet-Viguier died on 19 September 1736. He had been the mission's wide-eyed beginner: the willing enthusiast whose grandfather had worked on the Cassini surveys of France and whose astronomer uncle had explored South America. He was, wrote La Condamine later, 'the most robust' in the team. He had so much to prove.

The sudden absence of young Couplet left a gap in the ranks. It was a difficult time for Louis Godin, whose close friendship with Couplet's explorer uncle had opened the door to the teenager's inclusion on the expedition. Five months would pass before Godin picked up a pen and wrote to Maurepas with the sad news. It was a low point in the mission's bumpy journey towards geodesy. Before the survey had started, the French team had been reduced in manpower by one tenth. Two lives, one and a half years and thousands of livres and pesos had been spent and they did not even have a measured baseline. In the weeks following Couplet's death, the three Academicians devoted a fierce energy to fixing the first side of the first triangle.

Each end of the baseline at Yaruquí was marked with a buried millstone beneath a tall, timber pyramid that would act as a signal. Between the two pyramids, they erected a line of marker posts then cleared a physical line on the ground. Local labourers were conscripted to help with the heavy lifting. Slots were knocked in walls, gullies were filled with soil and stones, trees were felled and scrub cleared. By the end of the month, a ruler-straight scratch cut for 7 miles along the plateau.

While this preparatory work was going on, La Condamine volunteered for an entirely different task, far removed from Yaruquí. Accompanied by the practically minded cartographer-engineer, Jean-Joseph Verguin, he re-ascended Pichincha's precipitous heights and erected a stone marker on the summit, a place, observed La Condamine with mild relish, 'considered inaccessible'. The summit marker was painted white and visible from both ends of the Yaruquí baseline. It became the first of the signals that would be used to create the chain of triangles that would stretch all the way south to Cuenca. While they were up there, the two men supervised the construction of a small hut that could be used as a shelter during the forthcoming observations.

La Condamine was back at Yaruquí by 28 September, ready to participate in the first measurements. In a decision that announced the scientific and physical rigour that would characterize the coming years, Godin and Bouguer agreed to measure the baseline twice. One team would begin measuring the line from the north, and the other team from the south. They would pass at the centre point and continue to the far ends, then both teams would compare their results. It was an effective method of double-checking the precise length of the baseline. And it meant that the two individuals most likely to cause the mission to self-destruct would not have to work together.

Each team was equipped with three copper-tipped, twenty-foot poles that had to be laid end-to-end, repeatedly. Trestles and cords were used to keep the poles level and straight. The effect of humidity and heat on the lengths of the poles was monitored by regular comparisons with the one-*toise* iron rod they had brought from France to use as a standard unit. On the tilted surface of the plateau, it felt as if the sun was bouncing between the facing cordilleras, creating an asphyxiating furnace. Sudden whirlwinds filled the air with choking sand. On one occasion, recorded Ulloa, 'an Indian, being caught in the centre of one of these blasts, died on the spot'. Every midday, the teams had to withdraw into their tents, where they cowered until the temperature dropped and they could resume measuring. Initially, they were able to manage no more than 250 feet a day. In mid-October the two teams passed at the midpoint, and by the beginning of November they had both reached the far ends of the baseline.

At the Oyambaro hacienda close to the higher, southern end of the baseline, the teams began working on the dense clusters of figures in their pocketbooks. On comparing their data, they found that the two results differed by less than three inches over 7 miles.

The final stage of the process was to remove from the measurements the numerical deviations introduced by the sloping terrain. By adjusting the figures, they were able to derive a definitive total for the baseline length, as if the line were running horizontally from its lowest point. The final figure, of 6,274 *toises*, or 7.59 miles, fixed in space the first side of the first triangle. At last, they had started the triangulation. Next, they would lay out the chain of triangles, over more than 200 miles. That, however, would have to wait until La Condamine had undertaken his round trip to Lima for funds. In the meantime, the mission would base itself in Quito. On 5 December, they left the hot tabletop of Yaruquí and rode up towards the cooler airs of the city.

The *tremblement de terre* of 5 December lasted for forty-five seconds. Ten leagues south of Quito, buildings collapsed and several people were killed. The earthquake was an unnerving reminder of the mission's own unstable geology.

A letter had arrived for La Condamine while they had been working at Oyambaro. One of the mule trains that made the long and arduous traverse of the Andes that season was bearing the first package to reach the mission from France. The package had taken over a year to travel from Paris to Peru. In it was the letter Maupertuis had written back in September 1735 to his friend La Condamine, announcing that he was leading a geodesic expedition to the Arctic Circle, where it would measure the length of one degree of latitude 'so that nothing shall lack in the determination of the Figure of the Earth'.

The news from Paris sent tremors through the unstable nucleus of scientists in Quito. Maupertuis had recruited the young, innovative mathematician Alexis-Claude Clairaut. Before sailing north, the two of them were going off 'to spend the holidays with M. Cassini to practise some practical astronomy'. Maupertuis wrote of setting off for the Gulf of Bothnia in March 1736.

Reading the letter in November 1736, La Condamine assumed that Maupertuis was back in Paris already, with evidence that settled the prolate/oblate debate. Isolated in Peru, the three Academicians knew that the first to produce a definitive figure would claim to have settled the shape of the Earth. Whichever expedition returned second would be corroborating the evidence. The news left the mission in uncertain territory. Adrift in South America, Godin, Bouguer and La Condamine had to decide whether to abandon their equatorial survey, or whether to press on and return to Paris with their own arc of the meridian.

They could not abort the mission. If Maupertuis had returned from the Arctic Circle without definitive figures, the Academy would be no closer to knowing the shape of the Earth. The *merde* heaped upon the Academy and upon France would be of Andean proportions. If Maupertuis had succeeded in measuring one degree of latitude in the far north, then the figures from the equatorial survey in Peru would still contribute to knowledge about the Earth's shape. And already, the incidental science accumulating in mission notebooks amounted to more than Maupertuis could achieve up north. In the thin, volatile airs of equatorial America, it was a moment that demanded unity of purpose.

Godin now argued that the mission should instead produce a figure for one degree of longitude. To achieve this, he suggested a new survey line stretching west from Quito to the coast. In so doing, claimed Godin, the mission would create a new international standard for longitude, which could be used for comparative purposes by scientists elsewhere on Earth. It was an ingenious means of returning to Paris with a measurement that Maupertuis could not provide.

Bouguer disagreed, pointing out that the mountains and rainforests between Quito and the coast were less than ideal survey terrain and that the accuracy of the longitude method was inferior. The two were at loggerheads, with Bouguer ready to abandon the entire project unless Godin changed his mind. In large part, this was a mathematical row about the relative merits of two methods of determining the figure of the Earth. La Condamine was a bystander, ill equipped to join a fray that revolved around mathematical formulae, but he did know from personal experience that a geodesic survey through the rainforests west of Quito would be near impossible. Godin refused to relent, and a provisional truce was arranged. The lat–long decision would be delayed until the two options had been reconnoitred and mapped. In the New Year, Godin would undertake a westward journey along the equator, to evaluate the potential for a measurement of one degree of longitude. Bouguer would head north and evaluate the scope for measuring one degree of latitude. La Condamine – who was about to head south to Lima – would investigate the terrain south of Quito, with a view to contributing to the latitude argument. Yet again, the Academicians were proving that consensus was more painful than dissent.

With every week that passed, tangential events in Quito pulled the simple practicalities of triangulation further out of line. For much of December, their attention was consumed by the forthcoming solstice, on the 21st. Only twice a year would they have the chance to observe the obliquity of the ecliptic: the angle of tilt of the equator relative to the plane of Earth's orbit around the sun. Much to Bouguer's frustration, they had missed the 21 June solstice because they had only just arrived in Quito and had been 'delayed by some obstacles', not least the lack of a functioning observatory. Six months later, they were ready to take advantage of 21 December, when the sun would be at its furthest south in the sky. It was an opportunity not afforded to Maupertuis, or indeed anyone in Europe. Until the mission, nobody had taken modern astronomical instruments to the equator and, as Bouguer observed with relish, it is 'only near the equator that the obliquity of the ecliptic can be observed with great precision'. They had brought from Europe an instrument 'intended especially for this use'.

The mission's twelve-foot zenith sector had been designed, fabricated and assembled by the English boffin George Graham. It was so large and sensitive that it required its own observatory. The main frame of Graham's sector was a twelve-foot metal bar. Fixed to the bar was a long telescope. Attached to the foot of the bar, at right angles, was a graduated limb. A thumbscrew pushing on the end of the limb allowed the operator to incline the bar and telescope until they aligned with the chosen star. A pointed plumb bob suspended from the top of the bar recorded on the graduated limb the inclination of the bar and telescope. The entire assembly had to be suspended from above in such a way that it pointed upward at the sky and could be moved in two planes. This was achieved by suspending the sector from a pivot attached to a beam which had to be at least ten feet above the ground. A hole in the roof allowed the telescope to view the night sky. To take star sights, the operator had to recline backwards on the floor, with his face pressed against the adjustable eyepiece of the telescope. Observations, of course, were only possible in cloud-free conditions, at night.

With the zenith sector set up in the Santa Barbara observatory, Godin, Bouguer and La Condamine watched the ecliptic approach. On the 19th and 20th, the sky was overcast, but on the 21st, they managed to observe the angle. They stayed with the zenith sector through till the 27th, then compiled a report for the Academy in Paris. The three Academicians now had a calendar for the next year: they would start the triangulation survey after observing the next solstice, in June. By then, La Condamine should be back in Quito with pesos.

While the Academicians had been confined to their observatory, the streets outside had grown taut with tension. The close of 1736 marked the end of Alsedo's term as president of the *audiencia* and triggered a realignment of power between Quito's two ruling factions. For the mission, it was an unhelpful development. Alsedo's vanity, brutality and corruption did not make him a particularly wholesome individual, but he had not impeded the mission's scientific work. He had been the head in Quito of the Spanish-born caste of *chapetónes* whose leading figures included friends of the mission, among them the city's Jesuits. Jorge Juan and Ulloa had rooms in the home of Valparda y la Ormaza, accountant of the royal treasury, crown attorney and one of the most influential members of Alsedo's *chapetóne* faction. The incoming president looked like trouble. José de Araujo y Río was the native-born scion of Lima merchants who had deployed his legal training and avarice to buy the presidency of Quito for 26,000 pesos. It was a lot to pay, but the presidency in Quito allowed Araujo to connect a network of family and friends who controlled the Spanish ports of entry to South America. He rode into Quito on 26 December at the head of a mule train bearing 130 crates of contraband silks, porcelain, wines and silver. For Quito's native-born population, the change of president was long overdue. Araujo was unimpressed by the presence of the French scientists, an antipathy reciprocated by the mission.

As the year turned to 1737, the mission was in a precarious state, undermined by a rival expedition to the Arctic Circle and by an inability to decide whether they had come all the way to Peru to measure longitude or latitude. The last vestiges of Godin's leadership were being carried away by Andean winds. Future funding depended upon La Condamine surviving a 4,000-mile round trip to Lima in order to exchange a piece of paper for pesos. The arrival in Quito of Araujo was an unfortunate complication. Suspicions that the fundraising sell-off of mission property involved smuggled goods were passed to the Viceroy in Lima. As La Condamine was readying to leave for Lima, charges against him were being prepared. And then one of the mission's Spanish lieutenants managed to kill the president's secretary.

9

The killing had been seeded two years earlier. Back in 1735, a consignment of scientific instruments had been sent from Paris to Cádiz for loading on to the two Spanish warships sailing to rendezvous with the mission in Cartagena de Indias. But the four crates did not arrive at Cádiz in time. When the instruments eventually caught up with the mission in Quito, an outstanding charge of 20 pesos was demanded to cover mule transport from Guayaquil. Grabbing his first opportunity to frustrate the visiting scientists, Quito's new president instructed his treasurer not to pay the delivery charge and impounded the instruments.

The reaction was swift and it came from one of the mission's hotheads. Living under the same roof, Jorge Juan, Ulloa and crown attorney Valparda each had reason to challenge Araujo's actions. Valparda belonged to Quito's old-guard *chapetóne* faction and was no friend of the new *criollo* president. Jorge Juan and Ulloa saw a corrupt official impeding a royal directive. By impounding the scientific instruments, the president of Quito's *audiencia* had contradicted orders that came from the King of Spain to 'aid and assist' the mission, 'to show them all friendship and civility, and to see that no persons exacted of them for their carriages or labour more than the current price'. The 20 pesos was both a rip-off and a contravention of the king's instructions.

Ulloa's opening thrust was executed with a pen. He wrote to the president addressing him as *Vuestra Merced*, 'Your Grace', a term reserved for equals rather than superiors. Predictably, Araujo was insulted and returned the letter to Ulloa accompanied by the demand that the young Spanish lieutenant address him respectfully, as *Señoría*, 'Your Lordship'. The red mist descended and Ulloa forced his way into the president's quarters and subjected Araujo to a blistering tirade that could be heard across the plaza. Humiliated, Araujo ordered the arrest of Ulloa. Next day, a unit of armed militiamen intercepted Ulloa and Jorge Juan in the square. Ulloa was slammed against a wall and Araujo's secretary made the fatal mistake of drawing a pistol on the two most dangerous men in Quito. With his flintlock in one hand and his sword in the other, Jorge Juan dispersed the militiamen, wounding two. One of them was the president's secretary. He died slowly.

The killing of the president's secretary by a member of the mission was unhelpful to the future of equatorial science. The two lieutenants sought sanctuary in the Jesuit seminary. After lying low for a week, Jorge Juan effected a tactical withdrawal through a small window at 2 a.m. and rode out of Quito, heading south.

The success – or failure – of the Geodesic Mission to the Equator now resided with a mercurial individual riding away from the equator.

La Condamine had decided not to take the normal route. Instead of the well-travelled mule-trail over the mountains to Guayaquil, then a sea voyage along the coast to Lima's port of Callao, he chose to ride overland all the way. He was the first recorded Frenchman to have attempted the 2,000-mile trek along the Andes and he was equipped with a selection of books and his compass, quadrant, pendulum and barometer. Not recorded in his memoir were the identities of his companions, who must have included his servant and a guide. The overland journey promised plenty of interest. For the first 200 miles, he would be riding along the proposed route of the geodesic survey. La Condamine would be able to see for himself the nature of the terrain. Of particular interest would be the potentially troublesome mountains south of Riobamba and the landscapes around Cuenca, where the mission anticipated establishing its southern baseline. They would need a level plain similar to the one used for the northern baseline at Yaruquí. But La Condamine – as ever – was also planning extracurricular research. En route, he would 'observe the latitudes of remarkable places', draw maps and pursue his fascination with the lost world of the Incas. And he would search for the source of Jesuit's powder.

The total distance to Lima was reckoned to be 400 leagues. Initially, La Condamine was able to cover more than 10 leagues a day, but south of Riobamba, where the mountains crowded together and the road was seldom level, he soon found that 7 leagues was 'a strong day'.

Several days' riding beyond Cuenca, his little troupe reached the town of Loja. Two leagues south of the town, in forested hills climbing towards the western cordillera, La Condamine was able at last to gaze upon a cinchona tree. He had with him a set of notes prepared by the mission's doctor, Joseph de Jussieu. For three days, the urgent trek to Lima was delayed while La Condamine explored the stands of cinchona, filling his notebook with descriptions and drawings.

As they hurried south between the mountains, there were *tambos* – former royal hostels – to see and the tumbled walls of what might have been fortresses or temples. This section of the Andes marked the most northern extent of the vanished Inca empire. The ruins here were nothing like as stupendous as those reported to exist at Cusco, far to the south, but there was much to interest an acquisitive French collector. La Condamine's bags swelled with artefacts: 'some precious works of art from the ancient Peruvians, & various curiosities of natural history . . . several small idols of silver and a cylindrical vase of the same material, about 8 or 9 inches tall and more than three across, with masks chiselled in relief . . .' This exquisite silver piece was 'as thin as two sheets of paper held together, and the sides . . . were grafted squarely onto the base, at right angles, without any trace of solder'. For La Condamine, these were the links to a world that had been wiped from the Peruvian landscape by the colonizing cavalries of a global superpower. He longed to keep riding, from Lima on to the ancient capital of the Incas, Cusco, high in the mountains to the south-east, but '180 leagues of bad roads' forced him to 'renounce this project'.

Six weeks after leaving Quito, La Condamine rode into Lima. Armed with his passport and various letters of introduction, he managed to blag his way into the Viceroy's palace. He stayed for nearly three months. Securing the mission's funds proved more difficult than expected. His initial contact was unable to exchange the letters of credit because virtually all of Lima's cash had been dispatched by galleon to Panama. La Condamine had to resort to dirty money. In Lima at the time was the Panama agent of the South Sea Company. Thomas Blechynden had come to the city to collect debts and was eager to exchange his cash for letters of credit. La Condamine received 12,000 pesos and Blechynden left with credits for 60,000 livres that he could trade in Cádiz or Paris. The terms, noted La Condamine, 'weren't onerous'. Blechynden was a seasoned slaver and wheeler-dealer. Ten years earlier, he had been hauled before the company's court of directors for smuggling slaves through Portobelo. Many of the South Sea Company's agents were engaged in robbery and smuggling and Blechynden may have been using La Condamine to facilitate a money-laundering exercise. From the Englishman's point of view, the exchange was a sensible precaution: he was a long way from home and letters of credit were less of a health risk than a huge stash of cash.

Besides the Blechynden deal, La Condamine tried to top up the mission's geodesy chest with a request to the Viceroy for additional credit. Villagarcía had already rejected Godin's approach, but La Condamine came bearing a letter from the Duchess of Saint-Pierre, whose family had diplomatic connections with Spain. The Viceroy agreed to forward the matter to the Council of Finance. Although La Condamine eventually succeeded in winning approval for 4,000 pesos of credit, the complicated process led him down yet another cul de sac of distraction, which would end up costing the mission far more than 4,000 pesos in time and trouble. La Condamine took the financial application as a personal test, warming to the idea that the Council of Finance met only for 'extraordinary cases' and that the interminable representations and legal drafts caused him to undertake his 'first apprenticeship in the profession of solicitor'. From this point on, La Condamine could add 'lawyer' to a colourful CV that already noted him as a soldier, scientist, explorer and anthropologist. The list would get longer. And so would the distractions.

While in Lima, La Condamine packed a trunk with items he had collected since leaving Quito: the idols and vases, shells and balms, 'a dictionary and grammar of the Inca language' and 'a molar tooth, petrified in agate, weighing two pounds'. It was a gamble, but the routes from Lima were less hazardous than those out of Quito. A Spanish frigate was due to sail from Callao on 1 May, loaded with 'funds from the galleons of 1732'. With such a valuable cargo, the trunk would be well protected as far as Panama, where a British factor would forward it across the isthmus to Cartagena de Indias, where – hopefully – the trunk would be safely stowed in a Spanish ship bound for Europe. La Condamine addressed the trunk to the French consul in Cádiz.

La Condamine's stay in Lima was extended by a couple of surprises. Firstly, he learned that he had been accused of smuggling by the president of Quito's *audiencia*, who provided evidence that the French scientist had turned the Jesuit seminary into a shop for illicit trading. Customers had been observed coming and going 'at undue hours'. La Condamine was also accused of absconding to Lima 'loaded with prohibited goods', a charge that led to his room being descended upon by one of Lima's criminal judges, who demanded his keys and made a full inventory of the Academician's belongings. A report on the search was sent to Quito.

The second surprise was the unexpected appearance in Lima of Jorge Juan, bearing news that he had killed Araujo's secretary. Contrition was not a regular visitor to the mission's various memoirs and this latest debacle was described by La Condamine as 'Jorge Juan's personal affair with the President'. Forming a united front, the French scientist and the Spanish lieutenant appealed to the Viceroy for his understanding in the matter. Fortunately for the mission, Jorge Juan and Villagarcía were on good terms, having formed a helpful acquaintance with each other during their shared trans-Atlantic voyage on *Nuevo Conquistador*. Later, when Ulloa came to write up his account of the mission's activities, he skated over the killing, merely noting that Jorge Juan had gone to Quito 'in order to confer with the viceroy of Peru, for amicably determining some differences which had arisen with the new president'. With the backing of the Viceroy to complete the survey, La Condamine and Jorge Juan took a ship from Callao to Guayaquil, arriving in Quito on 20 June, the day before the solstice.

The reappearance of La Condamine and Jorge Juan with funding reinvigorated the mission. Even Jussieu was lifted. During the suspension of geodesy, he had lost his friend, the surgeon Seniergues, who had set off overland for Cartagena de Indias, where he intended to make money from private medicine. The melancholy Jussieu listened attentively to La Condamine's accounts of the cinchona trees in Loja, then wrote a four-page report on his colleague's findings.

While La Condamine and Jorge Juan had been away in Lima, leadership of the mission had slipped further from Godin. Inexplicably, Godin's determination to measure one degree of longitude had evaporated and his mapping survey of the equator west of Quito had not happened. Much later, Bouguer and La Condamine learned that, in March, Godin had received a letter from Maurepas in Paris, instructing the mission to restrict its ambition to the measurement of one degree of latitude. Godin had not mentioned the letter to Bouguer. A surprised La Condamine wondered whether the volte-face had been due to 'the orders he received' or because he had 'already changed his mind' and was now supportive of measuring one degree of latitude.

Momentum had been sustained by Bouguer and the mission's engineer-cartographer, Jean-Joseph Verguin.

Bouguer had toiled to and fro on the dirt tracks north of Quito, creating a detailed map of the terrain that would form the first part of the geodesic survey. Among his aims was the vital task of identifying the corners of each triangle. Every corner, or vertex, had to be inter-visible with the corners of adjacent triangles. When Bouguer returned to Quito in May, Verguin set off southward to create a corresponding map of the survey points between Quito and Riobamba. By the time he returned in June, the northern half of the survey route had been mapped. These two cartographic excursions were vital preparation for the coming season of surveying.

Two years after leaving Europe, the scientists were ready to begin the survey that would determine the shape of the Earth. At last, they were on the equator with the funds, instruments, maps and the seasoned personnel required to establish and measure a 200-mile chain of triangles spanning more than three degrees of latitude.

Triangulated surveys were old geodesy. If the length of one side and two angles of a triangle were known, the length of the remaining two sides could be computed. By adding successive triangles to the first, entire regions – or even countries – could be mapped by triangulation. The first printed treatise on triangulated surveys had been pulled from a Flemish press in 1533. In it, the globe-maker Gemma Frisius had described how to establish a measured baseline for the first triangle and how the selection of visible vertices such as church towers would allow the surveyor to measure the angles between the vertices of successive triangles. The Cassinis were criss-crossing France with chains of triangles that would be used to create the first, detailed, accurate national map. Instead of church towers, the Geodesic Mission to the Equator intended to use mountains for their vertices.

In execution, a triangulated survey required little more than repetitive accuracy, but translated to the Andes, the challenges were formidable. Starting outside Quito on the Yaruquí baseline, the mission's surveyors would work southwards along the corridor between the eastern and western cordilleras, using signals placed on high points as the vertices of the triangles. At each signal, they would set up a station and observe the angle between the other two signals of the triangle. They would be working in extremes of heat and cold over terrain that was frequently roadless and populated by communities who might prove unhelpful. On the higher observation points, cloud cover would be a constant problem. They all knew that this would be a survey of two halves: from Quito to the halfway mark near the town of Riobamba, the chain of triangles could be fitted into the long, broad plain between the two parallel cordilleras, using the peaks as the vertices of the triangles. But south of Riobamba, the cordilleras snapped together and the natural corridor was filled with a confusing congregation of gigantic massifs. The second half of the survey would be conducted in extremely difficult terrain, with remote vertices and appalling communications. The preparatory maps suggested that thirty triangles would be needed to span the arc of the meridian between Quito and Cuenca.

The three French Academicians had agreed on the method. Repeating the system of double-checking adopted for the baseline survey at Yaruquí, they would divide into two survey groups headed by the most able scientists and observers. Godin would be accompanied by Jorge Juan. Bouguer would be accompanied by La Condamine and Ulloa. Each survey group would have a dedicated support team who would set up the signals in advance and organize the transport of tents, food and instruments. Hugo and Godin des Odonais were tasked to accompany Godin, while Verguin and Bouguer's servant, Grangier, would work with Bouguer's group. The backbreaking physical graft would be carried out – as ever – by the mission's anonymous cohort of servants, slaves and local porters. The continued absence of Seniergues in Cartagena de Indias meant that the mission's health would rest mainly upon the shoulders of the suffering Jussieu.

By early August, final preparations were being made. The quadrants were checked again. Compasses, telescopes, maps and notebooks were collected. Tents were packed. Food was bought. Muleteers were sought. On 14 August, Bouguer's team left Quito, heading for the signal on top of Pichincha. One week later, Godin's team rode out of Quito bound for the signals at each end of the Yaruquí baseline and then for a signal on Pambamarca, a mountain in the eastern cordillera. The survey was on.

10

Antonio de Ulloa y de la Torre-Guiral lay unconscious on the mountainside. 'I fell down,' he wrote later, 'and remained a long time without sense or motion; and, as I was told, with all the appearances of death.'

The twenty-one-year-old naval lieutenant had set off from Quito with Bouguer, La Condamine, their servants and five porters, to climb to the hut and the signal that had been erected the previous year on the summit of Pichincha.

Mules had carried the surveyors and their equipment up the gentler terrain of the lower slopes, but as they gained height, the steep, rocky flank of the volcano had forced them to ascend on foot. La Condamine had been going well. He had climbed Pichincha before and was unafraid of its perils. Bouguer – who had suffered so much from sea-sickness – was climbing strongly. But Ulloa was struggling. He was a seafarer, not a mountaineer. The rough terrain and giddying exposure were unlike anything encountered at sea. And the cold air was so thin that he panted continuously for breath.

The decision was taken to split the group. While the servants and porters would remain with the instruments, Bouguer, La Condamine and Ulloa would continue towards the summit. There was no defined trail to follow and the trio scrambled upward over screes and jagged lava. With every step, breathing became more laboured. Precipitous drops threatened to snatch their footing. Then Ulloa collapsed.

To Bouguer and La Condamine, it was unthinkable that another member of the mission should appear to be on the edge of death. And Ulloa was vital to the success of the survey. His navigational skills and competence with instruments had qualified him as a core contributor to the mission's scientific aims. He was – alongside his slightly older compatriot Jorge Juan – the diplomatic go-between that made it possible for a gaggle of French scientists to operate in a Spanish colony. Ulloa was half of the mission's close-protection team. He spoke the language. He was a swordsman and a marksman.

When Ulloa regained consciousness, it was clear to his companions that he could climb no further. With difficulty, he descended the rocks to rejoin the servants and porters on easier ground, where he spent the night in a cave, rehydrating and resting. Next morning, he felt able to stand. Again, he struggled on the volcano's cliffs but, with help, managed to ascend to the summit.

Pichincha's peak was like the point of a needle. La Condamine knew what to expect, but Ulloa was appalled: 'Our first scheme for shelter and lodging, in these uncomfortable regions, was, to pitch a field-tent ... but on Pichincha this could not be done, from the narrowness of the summit; and we were obliged to be contented with a hut, so small, that we could hardly all creep into it.' In this tiny, rudimentary shack of timber and hides, the men would have to live while waiting for a window in the weather. A small tent was erected in a gap between rocks and into this crammed the five servants. The hired porters opted to sleep lower on the mountain, in the cave.

The observations could have been completed in one day, had the skies cleared. But the microclimate clamped to the volcano's cliffs generated unpredictable onslaughts of wind, snow and mist. Bouguer wrote of being:

> continually in the clouds, which absolutely veiled from our sight every thing but the point of rock upon which we were stationed. Sometimes the sky would change three or four times in the space of half an hour; a tempest was followed by fine weather, and in an instant after, thunder, loud in degree to its proximity, struck upon our ears; our rock producing the same effect with regard to it, as the sands of the sea when the waves dash against them.

Young Ulloa, who had survived countless storms at sea, found himself mesmerized by the mountain weather:

> When the fog cleared up, the clouds, by their gravity, moved nearer to the surface of the earth, and on all sides surrounded the mountain to a vast distance, representing the sea, with our rock like an island in the centre of it. When this happened, we heard the horrid noises of the tempests, which then discharged themselves on Quito and the neighbouring country. We saw the lightnings issue from the clouds, and heard the thunders roll far beneath us; and whilst the lower parts

were involved in tempests of thunder and rain, we enjoyed a delightful serenity; the wind was abated, the sky clear, and the enlivening rays of the sun moderated the severity of the cold. But our circumstances were very different when the clouds rose; their thickness rendered respiration difficult; the snow and hail fell continually, and the wind returned with all its violence; so that it was impossible entirely to overcome the fears of being, together with our hut, blown down the precipice on which edge it was built, or of being buried under it by the daily accumulation of ice and snow.

Inside the hut, the men lay on a floor of packed straw, struggling to stay warm. They subsisted on boiled rice with bits of meat or chicken. It was so cold that food froze solid unless balanced on top of a chafing dish of hot coals. For drinking water, they had to melt ice and snow. Each morning, the porters left their cave and climbed to the summit of the mountain, where they dug away the snow from the front of the hut, unfastened the leather thongs securing the double-hide door, and released the three Europeans from their smoke-filled hovel.

Whenever the weather allowed, Bouguer, La Condamine and Ulloa took short excursions from the hut, kneading the circulation back into their limbs. For amusement, they rolled great boulders off the cliffs and listened for the booming percussion of impacts far below. 'We generally kept within our hut,' remembered Ulloa, 'on account of the intenseness of the cold, the violence of the wind, and our being continually involved in so thick a fog, that an object at six or eight paces was hardly discernible'. While the wind thundered, Bouguer pored over his treatise on ship design. They updated their journals and wrote letters to Godin, letters that would take at least two days to travel the 8 leagues to the tent on top of Pambamarca. They took temperature readings and eventually succeeded, with 'every difficulty imaginable', in setting up and settling the delicate, adjustable pendulum. Bouguer noted that the length of pendulum that beat at one-second intervals was shorter by 'thirty-six hundredths of a line' than it had been on the coast. For La Condamine, the result was less significant than the enjoyable claim that the pendulum measurement had been achieved 'at the greatest height where it had ever been made'. Periodically, they peered into the swirling, freezing mists, hoping for a parting in the cloud that would prompt a dash for the quadrant. With every passing night, they became weaker. Inside the hut, the frigid gloom reverberated with coughing.

Had the three men in the hut been less weighted with racial prejudice, they might have learned a valuable lesson from their hired companions. In the 1730s, the effects of altitude on the human body were little understood in Europe. Above 8,000 feet, most humans begin to suffer from lack of oxygen. Headaches, nausea, dizziness, breathlessness and exhaustion become more extreme with every minute in the 'death zone'. As oxygen thins with altitude, the brain swells. Confusion and hallucinations are common. Unless the sufferer returns immediately to a lower altitude, cerebral oedema and pulmonary oedema can set in, followed by death. The hut on the summit of Pichincha stood at just over 15,400 feet. Local farmers and herders were used to sleeping in isolated hamlets and shelters at 9,000 feet and were already acclimatized to altitude. And they knew far better than Europeans that the debilitating effects of daytime activity at altitude could be ameliorated by descending each evening to sleep. The 'work high, sleep low' practice was exactly what the porters were doing by camping in their cave partway down Pichincha then climbing the peak each morning to help the Europeans. By sleeping on the summit, Bouguer, La Condamine and Ulloa were gradually killing themselves, and their servants.

With time, their feet swelled and became so tender that heat was unbearable and walking 'was attended with extreme pain'. Their hands were covered in chilblains. Their lips cracked so severely 'that every motion, in speaking, or the like, drew blood'. Day after day, they peered vainly into the gloom, praying for a window in the clouds. Whenever the mists did part on Pichincha, banks of cloud obscured one or both of the two signals they needed to view with the quadrant. On at least one occasion, they clearly saw through their telescopes the tiny white dot of the signal on Pambamarca, attended by Godin and Jorge Juan.

For the servants in the small tent on the summit, life was virtually unendurable. They began to vomit blood and to experience unspecified 'violent pains'. According to Ulloa, their hands and feet were 'so covered with chilblains, that they would rather have suffered themselves to have been killed than move'. After five days on Pichincha, the porters failed to show up at the hut for their excavation duties. Four of them had decided to descend the mountain, leaving behind one of their number, who eventually climbed to the summit and hacked clear the door of the reeking hut. The rescuer was dispatched to Quito with instructions to pass a letter to the city's *corregidor* requesting replacement porters. Two days after the replacements arrived at the summit of Pichincha, they too deserted. Eventually, a more humane solution took shape: an overseer in charge of four porters who were relieved of their mountain duties every fourth day.

By the third week, the situation had become desperate. Although they misunderstood the principal cause of their maladies, the three Europeans knew that they were approaching states of physical and mental infirmity that would make it impossible to descend the mountain. 'Monsieur Bouguer's health was impaired,' noted La Condamine, 'and he needed a rest.' After twenty-three days on the summit, the trio and their servants were helped down the volcano to the oxygen-rich sanctuary of Quito.

In science, there is no such thing as failure. The Pichincha episode was the experiment that provided the solution. Back in Quito, they came up with a new plan. They would relocate the Pichincha signal to a lower altitude, where there should be less cloud cover, less exposure to the cold, wind and snow and better access to the plain below.

Meanwhile, the other surveying team had enjoyed an easier start. They had spent less than a week at their first designated stations at the two ends of the Yaruquí baseline, then ridden on through El Quinche towards the wall of the eastern cordillera and Pambamarca, a wavelike series of rounded summits strewn with the tumbled blocks of ruined forts. Pambamarca was 2,300 feet lower than Pichincha and accessible by mule and, although it was cold and windy, they managed by 1 September to fill their notebooks with the angles they needed. Then things started to go wrong for them.

Back in El Quinche, Godin tried to re-engage porters for the next mountaintop station, over on the western cordillera, 16 miles north of Quito. But despite the lure of pesos, the locals were 'discouraged by their recent sufferings on Pambamarca'. They could see what was coming: the peak that Verguin had mapped as 'Tanlagua' looked as steep as its neighbour, Pichincha. Known today as Loma la Marca, Tanlagua was a pale, pointed marker on the road to the north. It was only a little over 10,000 feet in altitude, but millennia of weathering had eroded this old lava dome into a defiant spike. It was far too steep and unstable for an ascent with mules. All the able-bodied men in El Quinche suddenly disappeared and – remembered Ulloa – 'the joint endeavours of the *alcalde* and priest to discover them proved ineffectual'.

The people of El Quinche had reason to distrust Spaniards: 200 years earlier, when the conquistador captain Sebastián de Belalcázar found the town's men away fighting the Spanish, he ordered his soldiers to slaughter all the women and children they could find.

The revolt of El Quinche was a reminder that the Geodesic Mission to the Equator was an unviable human unit without the cooperation of local populations: the anonymous workforce who made themselves available to carry awkward, heavy loads to remote mountaintops where their lives were risked for European science. In a regime that rewarded those with colonial clout, the powerless were cheap tools. Jorge Juan reported with indignation that 'the lowest class of inhabitants' in El Quinche had 'left their habitations and absconded' rather than climb Tanlagua. For two days, Godin and Jorge Juan were delayed in El Quinche, unable to continue with their survey because they could not carry their instruments, tents, baggage and food without hired help. Eventually, the priest persuaded his reluctant sacristan, together with those employed in his church, to accompany the mission's loaded mules as far as a farmhouse at the foot of Tanlagua. They arrived on 5 September.

With the heavy *marquise*, baggage and instruments piled on their backs, the local porters began climbing on the 6th. Unencumbered by weight, Godin and Jorge Juan reached Tanlagua's exposed summit at the end of the day, exhausted and unaware that the porters had been unable to climb further than halfway. The two

Europeans and their servants spent the night under the stars with a 'hard frost coming on'. By morning, they had lost so much body heat that they were unable to regain full use of their limbs until they had descended to warmer air. Tanlagua was frustrating. The sky was clear but, through their telescope, Godin and Jorge Juan were unable to see the signals they needed to measure because they had 'either been blown down by the winds, or carried away by the Indian herdsmen'. They returned to Quito without their observed angles.

By mid-September, Bouguer, La Condamine and Ulloa were back on Pichincha, occupying a camp some 1,300 feet below the summit. Conditions were less severe, and access to Quito less difficult, but the clouds and tempests persisted. By 7 November, they had withdrawn again to Quito without the required observations.

In three months of fighting snowstorms, winds, cold and cloud, the two teams had not closed a single triangle of the thirty or so they had to measure between Quito and Cuenca. Far more time had been consumed with survival than with surveying.

They persevered. In December, Bouguer, La Condamine and Ulloa returned to Pichincha for a third time and succeeded in taking the necessary observations from the lower station. And on the 20th, Godin's team returned to the farm below Tanlagua and submitted themselves once again to the four-hour foot ascent. Seven days later, they too had measured the angles they needed. They were learning how to live with the mountains. Pichincha, Pambamarca and Tanlagua were key stations. The scientists had proved to themselves that high-altitude observations were feasible.

The allocation of pain was never going to be synchronized between the two teams. On Pichincha, Bouguer's team had taken a battering, while Godin's team had an easier time at lower altitudes. And Bouguer, La Condamine and Ulloa suffered again over the New Year. Between 20 December 1737 and 24 January 1738, they were stationed at the two signals that marked the ends of the Yaruquí baseline, but their observations were hindered by bad weather and by signals that had been blown to oblivion or recycled by locals. Each time a signal vanished, somebody had to return to the high point and reconstruct a pyramid of timber, then paint it white. The signal on Pambamarca had to be repaired or replaced seven times. Eventually, La Condamine erected a gigantic cairn from the stones of a fallen fortress, then topped it with a sturdy timber cross. On 26 January, Bouguer's team climbed Pambamarca to take fresh observations, but they were impeded by ice, snow and winds 'so violent that it was difficult to stand'. The blasts slamming Pambamarca made it virtually impossible to stabilize the quadrant. They were on Pambamarca till 8 February.

Between gales, Pambamarca's ghostly forts produced for Bouguer a memorable spectacle. The Breton hydrographer was amazed by a phenomenon that he was later credited with being the first to analyse. It occurred on mornings, when a 'very brilliant rising sun' projected his image on to a nearby bank of cloud. Of particular fascination was the rainbow-coloured halo around his head and the fact that others could see their own image, too, yet not those of the companions they were standing beside.

Next on Bouguer's list was Tanlagua. Where Godin and Jorge Juan had endured a mutiny, an arctic night without shelter and then a second ascent, Bouguer's team enjoyed fine weather and were on the summit for a single day. For Ulloa, however, it was another trial by vertigo. He conceded that Tanlagua was 'but small in comparison of others in this Cordillera', but the steepness of its flanks was a personal trial for the super-fit seafarer. Far below his feet, he could see the pale dots of roofs and the green chequerboard of fields. One slip would lead to a long fall and 'the greatest care [was] requisite in fixing the hands and feet close and firm'. On the descent, the least dangerous tactic was a bum-slide, though, he noted 'this must be done gently, lest, by celerity of motion, you tumble down the precipice'. This brisk encounter with Tanlagua marked a pause in surveying endeavours for Bouguer's team. Five months after setting off from Quito at the start of the geodesic survey, they had successfully completed the first of the triangles on Verguin's map. There were at least thirty more to measure.

While Bouguer, La Condamine and Ulloa took a month out, Godin's team was frustrated by the first triangle south of Quito. The least difficult of this triangle's three stations had been placed on a gentle rise beside the village of Guápulo, only four miles from Quito. They were able to commute daily from town to the *marquise*, where their servants guarded the instruments overnight. The other two points of this triangle were more troublesome. Twice, Godin and Jorge Juan had to climb the peak of Corazón in the western cordillera and Guamani in the eastern cordillera. On Guamani, the problem was the location of the signal, which proved to be hidden when viewed from the signal on Corazón. When they moved on to the next triangle and its signal on the mid-slopes of Cotopaxi, they found that the new signal on Guamani was hidden from sight. To get around the problem, they decided to erect an intermediate signal between the two mountains.

By now it was March 1738 and the two teams had been engaged on the survey for seven months, yet they were still in sight of Quito. At their current rate of progress, the survey would not take months, but years. Matters came to a head on a tapered plateau set between two incised valleys below the eastern cordillera. The 'Plain of Changallí' – a *changalli* was the typical apron worn by Andean women – was a comfy spot for a station, close to the town of Pintag and surrounded by gentle farmland. The signal had been allocated to Bouguer's team but had been erected the previous year by La Condamine, who had soaked it in lime to make it stand out on the plain. The weather was fine and the surveyors were able to sleep in a nearby farmhouse. However, none of these comforts was adequate compensation for the disappointing hours spent at their telescopes. The Changallí plateau had been chosen as the location of a key signal intended to be inter-visible with the signals on Pambamarca, Pichincha, Corazón and Cotopaxi. But 'some of the signals were wanting', reported Ulloa, 'having been blown down by the wind'. Surveying could not resume until assistants had re-ascended the peaks and replaced the signals.

That March, there was a meeting of minds between the two teams. Weeks had been lost through problems with the signals. The large, white-painted pyramids were too susceptible to dismantlement by the wind and by locals who saw the inexplicable structures as a convenient source of timber and rope. Wherever possible, they would abandon the pyramids and instead use the white, canvas army tents – the *canonnières* – as signals. The *canonnières* were easier to erect and to move than the pyramids and their presence at every signal was already necessary as a shelter for the surveyors and their assistants. It was a neat evolution of scientific practice.

With Easter approaching, Bouguer led his team south from the farmlands of Changallí towards the snows of Cotopaxi. Of all the peaks that lined the great valley leading south from Quito, Cotopaxi was the most spectacular. On clear days, it rose above its neighbours like a bright and shining triangle; a pyramid of snow and ice; an inviolable signal. In the setting sun, it appeared to be strewn with glowing embers. At least 19,000 feet in altitude, it was monstrously tall. No human had ever stepped upon its summit. The particular problem Cotopaxi posed to the mission was related less to height than to girth. Since it would be impossible to establish a signal on the peak's summit, observations would have to be achieved from its mid-levels. But the volcano's huge diameter would make it very difficult to find a single location that would be visible from all of the neighbouring signals. Bouguer established his signal on the north-western flank of the volcano, where he had a clear line of sight to the signals at Changallí and on Corazón.

Yet again, European geodesy was frustrated by Andean storms. Ulloa's recollection of their trial on Cotopaxi was characteristically vivid:

> This mountain we ascended the 21st of March, and on the 4th of April were obliged to return, after in vain endeavouring to finish our observations. For, not to mention our own sufferings, the frost and snow, together with the winds, which blew so violently that they seemed endeavouring to tear up that dreadful volcano by its roots, rendering the making of observations absolutely impracticable. Such is indeed the rigour of the climate, that the very beasts avoid it; nor could our mules be kept at the place where we, at first, ordered the Indians to take care of them; so that they were obliged to wander in search of a milder air, and sometimes to such a distance that we had often no small trouble in finding them.

They descended the slopes of Cotopaxi without their angles. A return to the mountain was now necessary. The continuous need to revisit stations and repeat observations was a drain on energy and finances. Again, the pesos were almost exhausted. And again, the mission was losing momentum. As it happened, the Cotopaxi difficulties coincided with the disappearance of local labour. With Easter looming, every family in the central valley would be unavailable for a fortnight while ceremonies and festivities took place. Reluctantly, both teams returned to Quito. It would be three months before they were measuring angles again.

It should have been a pleasant interlude. After eight months of high adventure, they all needed a break. In Quito, there were beds and food. Easter was a festival of processions and services. The tension between Araujo and the mission's Spanish lieutenants had dissipated. But in the shadows of their exuberance lurked the perennial problem of pesos and of leadership.

The cash that La Condamine had collected in Lima had been spent. The 4,000 pesos promised by the Viceroy had not been provided in full. In the short term, Maurepas had provided a financial lifeline by sending 4,000 pesos, but it was insufficient to complete the survey. So Godin turned to one of La Condamine's friends.

Back in June 1736, when Ramón Maldonado took La Condamine on a short excursion to see the route of the proposed Esmeraldas road from Quito to the coast, the Frenchman had been an enthusiastic supporter of the project. He had, after all, just spent several weeks struggling along the same route in canoes and on mud-logged forest trails. Two years on, Ramón's younger brother, Pedro Vicente, had become governor of Esmeraldas province. In April of 1738, he was in Quito to share his road-building plans with the president of the *audiencia*. Unsurprisingly, he ran into members of the mission. They were hard to miss. With a generosity typical of the Maldonado family, he offered to lend Godin sufficient pesos to fund the mission on the next leg of its survey. For the French scientists, it was another extraordinary stroke of good fortune.

The latest financial crisis had been averted. But eight months of surveying in different teams had broadened the chasm between Godin and Bouguer. A letter written in May 1738 by Bouguer and La Condamine to the English astronomer Edmond Halley complained that Godin had not shared his findings on the obliquity of the ecliptic. Discord reached a new intensity. Somebody on the mission described the dire state of affairs in a letter to Maupertuis. On receipt of the letter many months later, Maupertuis vented his frustration to the Swedish astronomer Anders Celsius, describing the mounting strife between the two Academicians in Peru and the damage it was doing to the mission's objective. Maupertuis told Celsius of 'dissensions' and 'terrible scenes' and of a complete breakdown in communication that had already lasted six months. He worried that 'one of them will die and the 2 others will return, one from the east and the other from the west without having done anything. It is almost to be feared that they will cut their throats.'

There was indeed a knifing over the Easter interlude. During an altercation with a local, Bouguer's slave was fatally wounded. He had endured harsher hardships than Bouguer, humping unwieldy loads through humid coastal forests, enduring countless nights on icy volcanoes and running innumerable errands to villages and towns. His name failed to make the memoirs.

In July, the engineer, Jean-Joseph Verguin, produced an updated map of the mission's progress. Much of the map was blank. Set upon a neat grid of squares were the two parts of Peru familiar to the mission: on the left of the map was the coast from the Esmeraldas estuary to Guayaquil and, on the right, an incompletely drawn spine of mountains. Crossing the map at zero degrees was a thin red line that Verguin had used to mark the equator. And hanging as if by a nail from a point just north of the red line, marked 'Sig.l de Tanlagua', was a pattern of triangles. The triangles that had been measured by Bouguer and La Condamine were shown in a solid black line, while those measured by Godin were shown in pecked lines. Verguin had used tiny red symbols to mark towns, villages and farms. The mountains were marked as tinted, stippled mounds. Blue rivers cut through striped farmland. The map was a work in progress. The triangles reached only as far as Cotopaxi and a signal labelled 'Milin', some way short of the line Verguin had used to show the latitude of one degree south. They were less than one third of the way to completing the chain of triangles.

The science resumed with a bang. On 10 July 1738, a large cannon was fired five times from the peak of Panecillo, on the edge of Quito. The first three shots were aimed northwards, over the roofs of the city, towards the spot where Godin and Jorge Juan stood watching from the far end of the plain by the royal road to Guayl-labamba. The fourth shot was fired south-westwards over the plain towards a hacienda occupied by Bouguer and Ulloa. The fifth was fired vertically upwards. Had the Academy in Paris known of the barrage, it might have viewed the booms above Quito as warning shots directed at three French scientists who had become increasingly distracted from the task they had been instructed to execute. The smoking cannon on Panecillo was the latest extracurricular investigation. Quito's scientists were attempting to measure the speed of sound.

Casting off from Rochefort and Cádiz had freed all members of the mission to explore the wonders of a wider, unfamiliar world. Anyone peeking into the private notebooks being carried to and fro between the cordilleras would have found jottings on Inca architecture, gravity, the Quechua language, disease, colonial justice, flora and fauna . . . the list of interests was infinite. The 10 July cannonade had its origins twenty years earlier in an English country village, where a clergyman called William Derham had measured the time it took for the sound of a shotgun blast to travel over a measured distance. Derham's experiments led him to compile a list of nineteen questions yet to be answered about the speed of sound. Number 13 queried whether the speed of sound changed with height above the Earth's surface. Number 18 asked whether the speed of sound was the same at different locations on Earth. The leading members of the mission were moved to address both questions. Twice, already, they had tried to record the number of seconds the sound of a detonation took to travel a measured distance. The previous year, while La Condamine and Jorge Juan were away in Lima, Godin and Bouguer had arranged for a cannon to be fired from the summit of Panecillo towards the peak of Pambamarca, a distance calculated to be between 19,300 and 19,400 *toises* – just over 23 miles. Equipped with a pendulum clock and a telescope, watchers beside La Condamine's timber cross on Pambamarca began counting as soon as they observed the muzzle flash of the cannon. But all they heard was the gentle sighing of an Andean breeze. They tried again at the end of August, while Godin and Jorge Juan were on Pambamarca at the beginning of the geodesic survey. This time, the wind was very gentle. The two men observed two muzzle flashes from Panecillo but listened in vain for the thumps of the detonations. They concluded that the sound had been deadened by undulations between the two vantage points.

For their third attempt, in July 1738, they abandoned the idea of recording sound over a long distance. The two listening stations they set up were at distances from Panecillo of only 5,736 *toises* and 6,820 *toises* – less than 10 miles. Over the shortened distance, the observations would have to be more accurate. At the listening station manned by Bouguer and Ulloa, the air was still. Godin and Jorge Juan recorded a wind speed of 2 *toises* per second, blowing towards Panecillo. This time, the report from the cannon was audible and the results revealed sound speeds of 174 *toises* per second and 178 *toises* per second, figures that were comparable with those achieved in England by Derham.

The experiment was of particular interest to the two Spanish naval officers. Later, Jorge Juan and Ulloa would devote a chapter of their book – *Observaciones Astronomicas y Phisicas* – to the speed of sound experiments, concluding that they would benefit geometry, geodesy and navigation. In particular, they speculated that knowledge of the speed of sound might prove invaluable during naval attacks and for measuring distance over wide, open spaces.

The speed of sound experiment broadcast to the entire population of Quito that the Geodesic Mission to the Equator was back in business. Next day, the two surveying teams were on their way south to the triangulation stations that would take the survey onward towards Riobamba and Cuenca and – they hoped – ultimate success.

Over the next four months, they proceeded with dogged efficiency, measuring triangle after triangle. The problems that had plagued the early triangles had been solved. They now had a method that worked. Wherever possible, signals and stations were placed to minimize the risks. Time and trouble were saved by the use of tents for signals instead of timber-and-fabric structures. With Couplet dead, young Jean-Baptiste Godin des Odonais had assumed greater responsibility as a general assistant to both teams, riding to and fro between the cordilleras preparing the signal stations and troubleshooting for the three Academicians. From his native American travelling companions, Jean-Baptiste was collecting Quechua words and gradually learning their language. The youth from the languid Cher had found a fresh field of dreams. Perhaps one day he would slip the hold of his cousin, mentor and leader: Louis Godin.

The location of each survey station became embedded in the memories of everyone who had laboured to reach the high point, set up the tents and instruments and then to record the angles of the next triangle. Corazón repelled Bouguer's team for nearly one month. The eroded stratovolcano was a key station in the western cordillera, functioning as the vertex for four adjacent triangles, with a signal that had to be placed high in order to deliver such a wide radius of inter-visibility. Two days after successfully completing observations on Corazón, Bouguer, La Condamine and Ulloa rode south-eastwards across the valley, past a farm that Verguin had marked on his sketch map as 'Ilitiou', nestled beneath the looming ramps of white-flanked Cotopaxi. Immediately to the east of Ilitiou, they climbed to a station they came to know as 'Papa-urco', Father Mountain. Today, it appears on modern topographic maps as Ilito, with the triangular symbol of a geodesic survey point. It was 'of middling height', remembered Ulloa, an 'easy mountain . . . a kind of resting-place between the two painful stations' of Corazón and Cotopaxi. They completed their observations in five days and then rode upwards, towards the snows of Cotopaxi. After their terrible experiences four months earlier, the scientists had reason to dread a return to the volcano. Their anxiety was well founded.

With questionable foresight, Bouguer and La Condamine sent Ulloa ahead to set up the tents and instruments. Following behind, the two Academicians found themselves trapped by failing light on the side of the mountain. They were too high for cow sheds and had to settle for a night in the open. La Condamine's taffeta cape was fashioned into a tent and pinned to the ground with hunting knives and the two men wrapped themselves in Bouguer's long coat, resting their heads on their leather saddles. After an interminable night chilled to the bones on frozen ground, followed by an agonizingly cold dawn, the pair discovered that their mules had wandered off into the fog. Rather than stay together, they split up. La Condamine eventually found his mule and caught sight of a porter carrying bread and the mission's tent poles and sent him back down the mountain to help Bouguer. It was a navigational cock-up rather than a disaster, but it must have reminded Bouguer in particular that they were operating in unforgiving terrain. La Condamine relished the adventure and wrote later that – once he had sent his mule and food back to Bouguer – his first concern had been to set up his quadrant so that he could 'take advantage of the fine weather'. But as soon as he focussed his telescope on the first signal, it disappeared. The white sheet that had been draped over the signal to make it visible from Cotopaxi had just been blown away or stolen. Two days were spent replacing the sheet, but the weather held and – reported La Condamine – 'we completed in four days a station which might have cost us a month's work'. The next station, on the 12,900-foot peak of Milín in the western cordillera, was knocked off in six days.

While Bouguer's team was ticking off triangles like a pendulum clock, Godin's troupe was staggering from one problem to another. Since leaving Quito on 11 July, they had returned to the eastern cordillera to set up the new intermediate signal whose absence had prevented successful observations from Cotopaxi. From there, they had ridden on towards the high station on Cotopaxi, but on the way up the mountain Jorge Juan's mule slipped and 'fell down a breach four or five *toises* deep'. The fate of the mule was not recorded, but the Spanish lieutenant emerged from the ravine 'without receiving the least hurt'. By 9 August, they had completed all the observations from Cotopaxi but were then faced with a frustrating return to the intermediate station of Papaurco to check one of the angles.

Papa-urco was a tipping point. The members of Godin's team were worn down by their recent trials and there were plausible excuses for a break. 'Here,' explained Ulloa, 'they for some time suspended their operations, being called to Quito on affairs of importance, relating to the French academicians.' Bills of exchange vital for funding the rest of the survey had arrived in Quito from France. They were payable to La Condamine, who agreed that they should be collected by Godin.

La Condamine seized the moment. With triangulation paused, he broke free on a side-trip to the wild side of the western cordillera in search of gold and a mythical mountain that might change the course of equatorial geodesy. The Academicians had known for a long time that they needed to remove the height differentials from the signals so that the entire chain of triangles could be mathematically reduced to a level plane, at sea level. La Condamine had become convinced that the immense isolated peak of Quilotoa beyond the western cordillera might be the missing link between the chain of triangles and the Pacific. But La Condamine had more than geodesy in mind. Far below the volcano, down in the rainforests at a location known as 'Tagualo', was a lost gold mine.

La Condamine's partner in this quest for gold and geodesy was a newcomer to the mission's antics. The Marquis of Maenza was a well-established enemy of Araujo, the new president in Quito. And he also happened to own land around Quilotoa. Maenza offered to build a summit shelter to protect the Academician and his instruments. Later, La Condamine reported that his efforts to observe the Pacific from the Andes had been thwarted 'by a setback that was all too ordinary': fog. His attempt to find the lost gold mine was also unsuccessful. To reach Tagualo, he had to descend 9,000 feet from the volcano into the rainforest. It was like searching for a fish in the Pacific. Tagualo was beyond reach in an impenetrable, unmapped land of ravines and trees. La Condamine remained coy about his treasure-hunting adventure, but the map that eventually appeared in his published account carried the location of Tagualo and, beside it, the note: 'Mine d'Or perdue', lost Gold Mine.

With no time to spare before triangulation resumed, La Condamine had to hurry back over the western cordillera. En route, he deviated from the trail to visit the fabled crater-lake of Quilotoa, whose water was reported to spout fire. Local people claimed that, shortly after the lake was formed, a 'vortex of flame' incinerated nearby sheep and caused the waters to 'boil for a month'. La Condamine found a wide, scree-encircled lake of 'greenish colour'. He concluded that it must be the 'funnel' of a volcano 'which, after having erupted in centuries past, still sometimes ignites'.

When he reached his next scheduled station on the eastern cordillera, he found that Godin had returned from Quito. And he had brought more than money.

The letters had been written by Maupertuis and Clairaut following their return the previous year from the Arctic Circle. As La Condamine sat in his tent at the station on Ouango-tassin (probably the peak known today as Señora Loma), he learned that the Arctic expedition had succeeded in measuring one degree of latitude and that their figure of 57,437 *toises* was so much more than the 57,060 *toises* measured in Paris that the Earth must indeed be flattened at the poles, and therefore oblate. Apparently, the Newtonians had been right. Removed from France by months of travel, the Academicians in Peru had been expecting the news for some time and had already aligned their reasons for continuing with their own mission. Without an equatorial figure for one degree of latitude, the true shape of the Earth could not be computed. For navigational purposes, it was essential to know the precise form of the curve. And among the doubts about the accuracy of the Arctic result was the fact that the length of Maupertuis's survey was only one third the length of the proposed chain of triangles in Peru. The news from France added resolve and urgency to the Academicians in Peru. They needed to finish the job then get back to the Louvre and present their superior findings.

By the time Godin was ready to resume surveying, Bouguer had streamlined the triangulation process. Instead of visiting all three corners of every triangle, and observing three sets of angles, both teams would now visit only two vertices and take two angles. It was an economy born of necessity. Unless they found a means of accelerating the survey, it would never be completed.

Through September, the two teams zigzagged south towards Riobamba. As the survey's halfway-mark, this town had always been a tantalizing milestone. Godin's run in to Riobamba was relatively easy. By the end of September, he was at a station on a vantage point beneath the eastern cordillera and it proved 'one of the most agreeable'. It was warm, the countryside had a 'cheerful aspect' and the town of Pillaro was so close that 'they wanted for nothing'. By contrast, Bouguer's team had a chilly few days on the side of Carihuairazo, a shattered volcanic caldera whose snowy shards topped 16,000 feet. Although Ulloa was feeling unwell, they managed to complete the observations in six days, but as they were preparing to leave on 29 September, the tent began to sway from side to side. Outside, the hired muleteers were trying to load baggage as an earthquake shook the flanks of Carihuairazo.

Southward, the topography became more complicated. Instead of a long, uneven corridor flanked by the eastern and western cordilleras, there appeared to be mountains in all directions. To ease the triangles into this confusing obstacle course, Godin and Bouguer were forced to place two signals very close together, one on a peak called Mulmul and the other on the much higher peak of Guayama (now known as Cerro Igualato). As the condor flew, only 8 miles separated the two summits. Their close proximity was a worrying precedent for the surveyors, who feared that such a small triangle would impair accuracy. Both teams converged on a 'cow house' occupying a gentle eminence situated between the two peaks. It was a typical high-altitude shack used by herders from the valley to sleep in while their cattle occupied upland grazing. For the surveyors, it was a convenient base and they were able to set out each morning to their respective stations: Bouguer's team heading up to Mulmul and Godin's team making the tougher climb up to Guayama.

For Ulloa, Mulmul was a peak too far. He had endured countless nights in tearing winds and blinding snowstorms but, on this lowly mountain, he finally succumbed to bacteria. The cow house became a sickbay. On 20 October, when the measurements were completed, Ulloa was carried down the mountain and onward to Riobamba, where he was 'confined . . . with a critical disease'. Without the help of their stalwart lieutenant, Bouguer and La Condamine spent a torrid month traipsing to and fro between Guayama and two new, intermediate signals to the east. They didn't arrive in Riobamba until 8 November.

None of them had expected that it would take so long

to reach the survey's midpoint at Riobamba. They had surveyed by triangulation the equivalent of one and a half degrees of latitude and they had done it at altitude in a terrain of extremes. But they were a very long way from closing the triangle that would mark the completion of three degrees. When Bouguer and La Condamine rode wearily into Riobamba on the 8th, they found that Godin and Jorge Juan had left town the previous day, bound again for Quito on a quest to liquidate La Condamine's bills of exchange. What might have been a morale-boosting reunion at the halfway mark was instead overcast with psychological jeopardy.

I I

By November 1738, the Geodesic Mission to the Equator had submitted to its own centrifugal forces and flown
into fragments. The mission was leaderless and peso-
less. Godin and Jorge Juan were absent in Quito. Ulloa
was ill in Riobamba. Jussieu was depressed. Seniergues
had not been seen for months. Bouguer and La Condamine were tired. The expedition's anonymous *domestiques*
were more exhausted than anybody. And, nearly four
years after leaving France, the mission was a long way
from delivering a figure that would define the shape of
the Earth.

Triangulating three degrees of latitude had become a
never-ending trek punctuated by discomfort and near-
death experiences. Most overnights involved extreme
camping. A cowhouse was a luxurious upgrade. Every
station demanded the same ritual: the wait for clear skies,
the adjustment of the quadrant, the search for the distant, pale dot of a signal, and then the painstaking
recording of angles. Only the scenery changed. A triangle was a triangle. And Riobamba was only halfway from
Quito to Cuenca. On projects of barely imaginable scale,
halfway points are moments of vulnerability. To reach
the 50 per cent mark of an extreme endeavour is an
achievement, but it is also a warning that the entirety of

everything accomplished has to be repeated in full. The only way of negotiating the dread of continuation is not to think of the end but of the next day. And in Riobamba that November, there was good reason not to dwell upon future difficulties. They had done the easy half.

For most of the way from Quito to Riobamba, the two survey teams had been helped in their travels and observations by the presence of the great valley that ran like a corridor between the western and eastern cordilleras of the Andes. It was a landscape designed for triangulation: two parallel mountain ranges for the placement of signals which – weather permitting – would always be in sight of at least two signals on the opposing range. The triangles fitted the space like a geometric jigsaw. All the two survey teams had to do was to ricochet back and forth between the ranges, fed by casual labour and supplies from the farmhouses and villages of the great dividing valley.

South of Riobamba, the space between the two mountain walls was congested by monstrous massifs that would block sight-lines and impede communication and supplies. Extreme cold and high winds would be certain at the high stations, and the trails between the stations would be contorted or non-existent, but always arduous. Among the Academicians, only La Condamine knew what to expect. Back in 1737, when he had travelled overland to Lima, he had ridden the long and winding Inca road from Riobamba to Cuenca and beyond. He had seen the storms on the mountains and knew that it was bad country for surveyors. The second half of the survey would make everything they had navigated so far look like a walk in the plaza.

Riobamba was the intermission between two acts. In almost every way, it was the perfect place for a bunch of knackered surveyors. Before the conquistadors brought conflict and disease to this compact, fertile plain below Chimborazo, there had been an ancient town here, cradled beside the river. Thirty minutes' walk to the south glittered a delightful lake. It had everything a town should want, apart from earthquakes. Native communities who survived the Spanish onslaught had been converted to Christianity by colonists who laid out a central plaza and a small town of regular streets. Riobamba's houses and public buildings were built in a mellow stone and some even had a raised storey. Of particular appeal to the mission was the town's cultural role as a haven for 'eminent families' who had developed the habit of migrating from their town houses to their country haciendas between December and June, when the cold winds blew down from the snows of Chimborazo. It was to the family hacienda of José Dávalos at Elén that the remnants of the two survey teams repaired to rest and to become entranced by Joseph's three talented daughters, the eldest of whom could paint, play half-a-dozen musical instruments and understand French. Unfortunately for La Condamine, she was intent on becoming a Carmelite nun. While they were at Elén, news came through from Quito that Godin had been 'seized with a fever, which [had] brought him very low'.

*

Eleven days after Bouguer and La Condamine lurched into Riobamba, bright skies lured them back to the mountains. It was a remarkably quick turnaround for them both and a measure of their determination not to let Godin's absence delay the survey any further. Putting geodesy ahead of health, Ulloa roused himself from convalescence and rode out of Riobamba with them.

They headed up to a signal that formed the vertex of four triangles whose completion would open the way southward towards Cuenca. It was a 13,000-feet vantage point they knew as Sisa-pongo or Dolomboc, high in the western cordillera. As ever, clear weather was essential and, for once, they were lucky. Not only were they able to capture the angles they needed, but on three occasions they measured the azimuth of the setting sun, which allowed them to check the overall orientation of the chain of triangles. While they were up there, they also enjoyed the spectacle of distant Sangay spouting flames into the night sky. Sangay lay far to the east, on the edge of the Andean cordillera, and its sudden transformation into a fiery gout turned it into a perfect geodesic signal. The two Academicians busied themselves with the quadrant, fixing the location and height of the erupting volcano.

After a productive week of observations, Bouguer's team returned to Riobamba, eager to take the triangles on to Cuenca and completion of the survey. But Godin and Jorge Juan had yet to return from Quito with the funds essential for further progress. Geodesy was stalled.

For Bouguer, the inability to triangulate was a

heaven-sent opportunity to play with his instruments. Experiments he had conducted on Pichincha and in Quito had provided valuable figures for the effects of altitude on gravity. Now, he wanted to test Newton's law of gravitational attraction by measuring how much a hanging plumb-line deflected towards the mass of a mountain. Back in October, he had diverted from triangulation to spend a week tripping over in the thickly vegetated environs of the volcano Tungurahua, searching in vain for a site close to the mountain that could be used to set up the sensitive instruments. Other, more accessible mountain masses had been ruled out. Cotopaxi's evident vulcanicity raised the possibility that it might be partially hollow. Pichincha was deemed unsuitable because its multiple peaks reduced its mass. That left Chimborazo, which was both massive and less than a day's ride from Riobamba.

La Condamine needed little prompting. Instruments were assembled. Together with the plumb line, they would need a quadrant, a clock and a pendulum. Learning from the pendulum difficulties on Pichincha, La Condamine decided to commission his own, portable mountain pendulum, encased in a specially constructed elongated box that would shield the fragile components from the effects of wind and protect them from damage during the ascent of the volcano. Again, Ulloa decided that he was well enough to help the two Academicians with the observations.

On 29 November, they left Riobamba with their servants, a tent and the boxed instruments, riding north to

Chimborazo and setting up a base in a farmhouse on the lower slopes. They spent the 30th dicing with death. For ten hours, they scrambled in and out of deep gullies, gradually gaining height through banks of snow and loose rock. For La Condamine in particular, it was a desperate struggle. They were trying to reach a vantage point known as the Perch of the Condor. Eventually, they scrabbled up to the exposed buttress on the mountain's permanent snowline and pitched the tent on a ridge between two ravines. It was a marginal campsite and the men were kept awake at night by the rumble and boom of massive avalanches careering down the mountain towards the tent, only to be deflected before impact into the ravines on each side. Ulloa relapsed and had to be helped down the mountain and back to Riobamba. Bouguer and La Condamine stayed on, struggling with the instruments at the Perch of the Condor and at a lower station they established for taking comparative observations. After an excursion that lasted more than three weeks, they returned to Riobamba three days before Christmas and learned that Godin and Jorge Juan were expected imminently.

Suddenly, there was a rush. La Condamine and Bouguer knew they had to capitalize on *l'attraction newtonienne* before the geodesic survey resumed. La Condamine burned the midnight oil and by the 23rd had managed to compress the gravitational findings into a letter to his Academy friend Charles du Fay. Bouguer pored over the numbers and was disappointed to discover that the effect of Chimborazo's mass on gravity was far less than he had expected. He concluded that the mountain's density must be less than that of Earth. He wrote a brisk report on the experiment and, on the 30th, dispatched it to the Academy in Paris. As 1738 trickled into 1739, thoughts turned to home. In his letters to France, La Condamine wrote of being back by the end of the year. Given the length of time that it took for a letter to travel between South America and Europe, he expected no answers to the letters he was sending.

Impatience turned to frustration. Godin and Jorge Juan did not appear. La Condamine was desperate to resume the survey. In a meadow outside town, he worked on the quadrant, making adjustments, resolving discrepancies in the various tables. He fiddled with the pendulum and revisited the gravity calculations. Meanwhile, Bouguer drifted away, La Condamine observing that his friend had 'retreated into the countryside near Riobamba, to make various observations, of which I had no knowledge'.

Yet again, the three leading lights had lost touch with each other. And yet again, there was a financial crisis.

La Condamine cracked. Godin and Jorge Juan had not showed. With every passing day, the wet season was seeping closer and, with it, the threat of cloud cover that would obscure the signals they needed in order to extend the triangles south to Cuenca. La Condamine decided to press on without Godin and Jorge Juan. As the only Academician with private funds, he dug deep again and provided Bouguer and Verguin with advances to cover the cost of mules, labour and food. They would split their three-man team into two. La Condamine would take over Godin's stations, leaving Bouguer and Ulloa to observe from the stations on their original list. The ever-dependable Verguin would travel ahead to erect the signals, leaving them in place so that Godin and Jorge Juan could use them when they eventually caught up. La Condamine rode out of town on 17 January, bound for the peak of Zagrún on the eastern side of the chain of triangles. Optimistically, he packed Graham's twelve-foot zenith sector, the instrument that would be used for the astronomical observations necessary to conclude their years in South America.

Good fortune was followed by bad. Zagrún was a prominent knoll on a long promontory poking north from the mass of peaks that filled the southern skyline. To reach the station, La Condamine had to cross a fero-cious river and then climb 2,000 feet to the exposed ridge. But the skies were clear and he concluded his observations in a mere three days. By the 21st, he had crossed to the western side of the chain of triangles and had slogged up to the next station, on the peak of Lalan-guso, where he had arranged to rendezvous with Bouguer and Ulloa. They were not there. And it was the bleakest of locations, on an exposed, treeless *páramo* at 14,000 feet, battered continuously by wind, rain and snow. The hired labourers cleared off, closely followed by a valet, who robbed La Condamine before leaving.

Bouguer and Ulloa eventually arrived at the Lalanguso station on the 25th. The appalling weather continued.

Gusts ripped two of the tents. Wind and rain pelted through the flapping canvas, exposing those inside to miseries of damp and cold. It wasn't until 31 January that they managed to complete the observations. Later, La Condamine would claim that Lalanguso provided 'one of the harshest of all our camps in the mountains'. Ulloa thought the next signal was even worse.

The continuing absence of Godin and Jorge Juan increased the strain on the remaining surveyors. On 2 February, Bouguer led a detachment towards the eastern side of the chain of triangles and a lofty, remote signal on a mountain known as Senegualap. Meanwhile, La Condamine and Ulloa headed down the western side of the chain for another station that should have been manned by Godin and Jorge Juan.

That night, La Condamine and Ulloa reached the town of Alausi, a tight-packed grid of streets overlooked by green, forbidding walls. Their station was hidden from sight to the north, on a peak described by La Condamine as 'conical, isolated and very steep ... about 1960 [*toises*] above sea level'. Verguin labelled it on his sketch map as 'Chougeay'; Ulloa, the Spanish-speaker, knew it as 'Chusay' (the peak that fits the descriptions most closely is 12,398-foot Cerro La Mira, used 200 years later as a geodesic survey point). La Condamine and Ulloa were not in good shape. Their tent was still wrecked from the battering on Lalanguso. And they were unsure whether Verguin had erected the next signal to the south. La Condamine had little choice but to ride on from Alausi for 3 leagues, find Verguin and then return to Ulloa. He caught up with his Spanish compatriot on the trail that climbed towards Chusay. Somewhere on the mountain's lower slopes they spent three nights having their tent repaired in 'an Indian cottage'. On the 6th, they slogged up to the summit and peered anxiously through the telescope, searching the serrated skyline to the south for the signal they expected. Then they searched south-eastwards. Where there should have been the bright, white dots of Verguin's signals, there was just an angular confusion of *páramo* and peak.

This was the problem they had been dreading. From Quito to Riobamba, the triangles had slotted neatly between the two cordilleras and the placing of signals had been relatively straightforward. Now, the tighter clutter of peaks made it extremely difficult to identify signal locations that were inter-visible. On his map, Verguin labelled this mountain region 'Assouaye'. The spelling varied, but the sound of the word was that of wind. 'The highest point of Assouaye', wrote La Condamine, 'seen from some distance, seemed to merge, projecting one on the other. From afar, only a mass can be seen . . .'

The weather on Chusay was atrocious. Three weeks after their first ascent, they were still waiting for the southern signals to appear. Communication with Bouguer was difficult. He was far to the east on a 6-mile whaleback of wind-buffeted *páramo* at an altitude of over 13,000 feet. He, too, was waiting for the missing signals. In an exchange of letters, Bouguer suggested that La Condamine leave Chusay and explore the onward mountains for feasible signal locations. But by then La Condamine was incapacitàted. On a day trip down the mountain to Alausi, his horse had fallen and bolted. La Condamine had managed to snatch his feet from the stirrups and kick himself clear, but his leg had been crushed. Lying on his bed with an immovable limb, he consoled himself that Chusay's endless rain and fog would have made observations impossible.

As soon as his injured leg could bear his weight, La Condamine set off on a search for the new signal locations. For eight days, he was 'wandering through the moors and marshes, finding no other shelter than caves dug in the rock'. He trekked between the peaks, one by one, drawing a map and eventually placing a signal on a summit he called Gnaoupan, which he could see connected all the necessary triangles. Unfortunately, 'a misunderstanding' led to Gnaoupan being ignored in favour of Sinasaguan, which was not in place for another month or so. On 21 March, six weeks after arriving at the summit of Chusay, La Condamine and Ulloa turned and descended without having taken their angles. Back again in Alausi, they found some familiar faces.

Louis Godin had returned from Quito with Jorge Juan. Seniergues was there, too, and Jussieu. And Morainville. And Verguin. Bouguer was still in the western mountains, but this was the most complete gathering of the mission for a very long time. It was a short, hectic reunion.

Seniergues had been away from the mission for nearly two years, making money from medicine in Cartagena de Indias. La Condamine – whose adventures on the French national lottery had funded his time in Peru – never failed to be impressed by the surgeon's ability to turn pain into pesos. In Cartagena, Seniergues had managed 'to secure by his industry a fortune'. He had also brought back from Cartagena a supply of glass barometer tubes for La Condamine, who shared them out with Godin and Bouguer. Seniergues was able to make a loan to Godin and to back an ambitious botanical field trip led by the mission's doctor-botanist, Jussieu.

Jussieu, Seniergues and Morainville had devised their own expedition. They were going south to the mountains around Loja, where they would research the cinchona tree, whose bark was such a valued treatment for malaria. They left Alausi on 22 March. At daybreak on the 24th, La Condamine and Ulloa were back on the heights of Chusay, enjoying a blinding sunrise and taking the last remaining angle they needed from that station. A couple of days later, on the 26th, the survey teams resumed their 'old order of march', with Bouguer observing with La Condamine and Ulloa, while Godin worked with Jorge Juan. Back in their tried-and-tested teams, they were optimistic that the last triangles would be closed in a few weeks.

For a month they trekked the plains and peaks. Tioloma was ticked off. They watched Sangay fill the night sky with rivers of fire. La Condamine had to undertake another solitary adventure, crossing the mountains for a fourth time to check the inter-visibility of onward signals. Then, on 17 April 1739, he found himself on the last of the high-altitude stations. Sinasaguan was a nightmare.

*

Booming gusts shook the fabric of the *marquise*. The poles shuddered and bent. Canvas sagged beneath the weight of snow. In the smaller, less robust *canonnières*, morale was low. Ulloa recorded the flight from the camp: 'The Indians who attended us, not willing to bear the severity of the cold, and disgusted with the frequent labour of clearing the tent from the snow, at the first ravages of the wind, deserted us.'

The signal on Sinasaguan had been carefully sited amid the jumbled uplands south of Alausi. It was one of the rare stations that brought both survey teams to the same spot, but it was a very exposed location, subject to frequent snowstorms. For the muleteers and their master, it was a brutal experience. They had taken shelter in a cave when the storm hit.

The tent was ripped down by the wind. When a replacement was erected, that too was destroyed. And so was a third. Stout tent poles snapped. Eventually, the surveyors had to seek shelter in a ravine. Their situation was worse even than it had been on Pichincha: 'While we were thus labouring,' wrote Ulloa, 'under a variety of difficulties from the wind, snow, frost, and the cold, which we here found more severe than in any other part; forsaken by our Indians, little or no provisions, a scarcity of fuel, and in a manner destitute of shelter . . .'

At the foot of the cordillera, the priest in Cañar had taken to offering prayers for the scientists, unseen in the 'blackness of the clouds' above the town. The foreigners had been given up for lost when, after a fortnight fighting the weather, they staggered down from the heights.

Passing through Cañar, they were viewed 'with astonishment' and received 'with the most cordial signs of delight, adding their congratulations, as if we had, amidst the most threatening dangers, obtained a glorious victory'.

And victory was almost within reach. The scruffy cavalcade that passed along the streets of Cañar on 9 May 1739 were on the road to the end of their survey. Bouguer, La Condamine and Ulloa could afford to dream of completion. The few remaining stations designated to their team were relatively low-lying and accessible. Godin and Jorge Juan were on a peak called Quinualoma in the wild east. But for Bouguer's crew, the end was in sight. Their next station was on the fabled peak of Buerán, home of the Cañari hill god. Old Taita Buerán (Father Buerán) was a small man in a greyish poncho who lived on the peak in a cave heaped with gold. When the mist slipped down the slopes, he would lure shepherds into his lair, where he would present them with precious metals that turned to animal dung as soon as the treasure was taken down to the lowlands of the humans. Taita Buerán was a familiar figure to Andean communities, a mountain god who promised betterment that proved chimerical.

Bouguer, La Condamine and Ulloa climbed into Buerán's mists on 10 May. It should have been an easy station. Buerán's slopes were not precipitous and the peak was only a couple of hours walk from Cañar, so the surveyors were able to indulge in a weekly commute from town. Besides 'the small height of the mountain,' wrote Ulloa, 'the town of Canar being only two leagues distant from it, we were in want of nothing. The temperature of the air was also much more mild than on the other deserts; besides, we had the great satisfaction of relieving our solitude by going to hear mass on Sundays, and other days of precept, in the town.'

Inevitably, the local Cañari became curious about the Europeans camping on old Father Buerán and one day Ulloa was accosted by 'a gentleman of Cuenca' who had seen him leaving the tent. The gentleman knew Ulloa's name and had heard of the mission but was utterly unable to comprehend why an intelligent European would be dressed 'in the garb of the Mestizos . . . the lowest class of people'. Nothing Ulloa said could convince the visitor that they were not perpetrating a hoax and that the story about measuring the Earth was a cover for mineral prospecting. Why else would Europeans subject themselves to 'such a dismal and uncouth life'? Ulloa, the disciplined young naval lieutenant, had become a bearded, grimy, smokey mountaineer.

There were 'terrible storms' on Buerán, storms that struck the Cañar region so badly that 'the animals, cottages, and Indians, suffered three times in a very melancholy manner by tempests of lightning'. For La Condamine, the tempests roared with a silver lining. After more than a week of continuous cloud cover on Buerán, he proposed to his companions that their collective skills could be better employed by descending the mountain and riding across the valley to the Inca ruins on the far hillside. It was an excursion that won La Condamine enduring admiration.

Wednesday, 20 May was spent exploring the passages, chambers and terraces of a ruin known locally as Inga Pirca, or 'walls of the Inca'. La Condamine had read widely on the conquest of the Incas but none of his sources – not Agustín de Zárate or Pedro de Cieza de León, or Francisco López de Gómara, or the Jesuit Father José de Acosta (who first documented altitude sickness), or even Garcilaso de la Vega, the great chronicler known as 'the Inca' – had left detailed descriptions or plans of major Inca monuments. La Condamine was breaking new ground. Here was a subject that demanded his aptitudes as historian and geographer, surveyor and field-worker. Two years earlier, while on his 'private voyage' from Quito to Lima, he had passed the tumbled stones of hostels and fortresses but had not had time to pause and explore because the demands of his mission 'permitted no delay'. An ambition to visit the ancient Inca capital of Cusco had been thwarted because it was too far off his route. Tumipampa's gold-sheathed palaces and emerald-encrusted walls had been obliterated by the sprawling grid of Spanish Cuenca. Of the few ruins he had managed to see, none compared in preservation to Ingapirca. That Wednesday, while Bouguer and La Condamine were recording the north side of the main platform, a local farmer was 'working at the demolition of what was best preserved there' and removing the Inca stones 'for a new building at the neighbouring farm'.

La Condamine was convinced that Ingapirca was not a palace but a fortress. As a military man, he recognized the worth of a naturally defended site. The ruins stood on a levelling between two deep ravines. Recalling how Garcilaso had described a room in the palace at Cusco that could hold 3,000 people, La Condamine was sure that Ingapirca's chambers were too small for massed ceremonial gatherings and that he was looking at the remains of a fortress built by the Cañari to defend their territory from the expanding empire of the Incas. As he had when visiting the Inca ruins at Callo, La Condamine marvelled at the precision of the stonework, each block fitting so neatly to the next that the 'joints of the Stones would be imperceptible if their exterior surface were flat, but this is worked in relief'.

For La Condamine, the visit to Ingapirca had been a revelation. A few days later, he returned to the ruins to take more measurements and to record with his compass the orientation of key structures. He also selected a suitable viewpoint and completed a sketch of the site.

Back on Buerán, the clouds eventually parted and Bouguer managed to snatch the angles they needed. They left the mountain on 1 June, heading for Cuenca.

That day, far to the west, Godin, Jorge Juan and their assistants were also looking towards Cuenca and completion. Quinualoma had been brutal. For three long weeks, they had been freezing their nuts off while waiting for the weather at one of the 'most disagreeable stations in the whole series'. They stumbled down the serpentine trail to the town of Azogues, dumped their instruments and baggage, and then took the road for Cuenca.

*

Cuenca became the rallying point for the scattered mission, and the grave of one of its members.

Just as he had with every town and city he had visited, Ulloa subjected Cuenca to geographical scrutiny and a written report. He began with the town's precise latitude and longitude and then a description of its location on a 'very spacious plain' watered by four rivers crossed by fords or bridges, depending upon the season. There were between 20,000 and 30,000 inhabitants, several convents, a Jesuit college, a couple of nunneries, three churches – the 'great church' for Spanish and those of mixed blood, San Blas and San Sebastián for the indigenous population – and a hospital that was 'almost in ruins'. Cuenca was, he decided, a town

> of the fourth order. Its streets are straight, and of a convenient breadth; the houses of unburnt bricks, tiled, and many of them have one storey, the owners, from a ridiculous affectation of grandeur, preferring elegance to security. The suburbs, inhabited by the Indians, are, as usual, mean and regular. Several streams of water, by great labour, are brought from the above rivers, and flow through the streets; so that the city is plentifully supplied; and for its admirable situation, and the fertility of its soil, it might be rendered the paradise, not only of the province of Quito, but of all Peru; few cities being capable to boast of so many advantages as concentrate here; but either from supineness or ignorance, they are far from being duly improved.

The average male in Cuenca exhibited the 'most

shameful indolence, which seems so natural to them, that they have a strange aversion to all kinds of work; the vulgar are also rude, vindictive, and in short, wicked in every sense'. Cuenca's women, on the other hand, were 'remarkable for an uncommon spirit of industry' and famous throughout Peru for their brightly coloured textiles and woven cotton, which they sold to traders and merchants passing along the Great Road. There is little doubt that Ulloa's view of Cuenca's men was tainted in large part by the forthcoming tragedy.

To Cuenca they came, in dribs and drabs. They had achieved an extraordinary feat. The chain of triangles stretched across more than three degrees of latitude, a distance three times greater than the Arctic triangulation managed by Maupertuis. Now they needed a southern baseline to match that of Yaruquí in the north. With a baseline at both ends of the chain, the accuracy of all the angular observations could be verified. But around Cuenca, level plains were rare. There were two possible locations. Seven miles south of town, the Great Road ran along the bed of a long, flat valley hemmed in by mountains. La Condamine had noticed the valley two years earlier, when he had undertaken his journey to Lima. He was convinced that the plain at Tarqui would be a suitable surface for the concluding baseline. Its only disadvantage was its distance from the southern end of the chain of triangles. More triangles would need to be added. There was an alternative plain, smaller and less even, but very close to Cuenca, at a place called Baños. Bouguer insisted on Tarqui. Godin insisted on Baños.

It would, of course, have been too simple if Godin and Bouguer had settled their old scores and agreed that one of these two options was the best. But the two Academicians were resolute in their opposition. It was a reminder that the mission's self-destructive instincts remained intact, despite the extraordinary vicissitudes its members had shared, and despite the fact that they were just a few weeks from completing their three-year geodesic survey. By duplicating baselines, they were doubling the measurements that remained. And prolonging their stay in Peru. And delaying the computation of the final figure.

And so it was that, as May turned to June 1739, the two warring 'leaders' gathered their troops to identify the end-points of the two new baselines. Only when these had been fixed could the chain of triangles be completed. Having erected their baseline signals, the two teams returned to the mountains to measure the concluding triangles.

The triangulation finale was tiring but trouble-free. For Bouguer, La Condamine and Ulloa, it meant a ride east from Cuenca down the valley of the River Paute to the small town of the same name, where they turned left up towards the 'Páramo of Yasuay', a vast, bleak massif with sides so vertiginous they had to climb to the 12,000-foot summit on foot, 'nor even by that method without great labour', recalled Ulloa. He reported that the temperatures on the exposed crest were 'far from being so intolerable as on Sinasaguan and the deserts north' and that they had 'cheerfully supported the inconveniences of this station'. They were up there for ten days. Ulloa had reason to be elated. It was his last serious mountaineering adventure. In mid-July, the little troupe carefully descended the mountain to the lush greenery of the valley floor, then followed the river back upstream, towards the next signal.

To link Yasuay to the baseline, they had to climb two more peaks. The first was Borma. Although it was a relative minnow at a little over 10,000 feet, the surveyors were anxious. Only if the Páramo of Yasuay was clear of clouds could they measure the vital angle between Yasuay and a signal that had been placed 11 miles north of Cuenca on a relatively accessible mountain spur called Cahuapata. In the event, they could not believe their luck: 'It was also our good fortune that Yasuay, contrary to our apprehensions, was clear and visible the whole 19th of July; so that we finished our observations in two days agreeably.'

Godin and Jorge Juan returned to Azogues, where they had left their instruments and baggage, and on 15 June scaled the heights of Yasuay. They descended on 11 July and added some small triangles, one of which used the tower of Cuenca's 'great church' as its vertex. By the end of July, Godin, Jorge Juan and Godin des Odonais had measured the length of the baseline at Baños and, by the end of August, Bouguer, La Condamine, Ulloa and Verguin had measured the slightly shorter baseline at Tarqui.

It was while they were plotting these final elements of the survey that the surgeon, Jean Seniergues, showed up in Cuenca, having left Jussieu and Morainville in Loja to continue researching the cinchona tree. In the four years since they had left Europe, all of the mission's members had changed in some degree. Each had been subjected to episodes of fear, illness and extreme discomfort among people they little understood. The effects of alienation affected everyone differently. Seniergues had become the performer of miracles, curing the sick and raising patients from the clutches of death. European doctors were few and far between in South America. In Cartagena de Indias, he had made big money and he had also cured the poor. Wherever he went, Jean Seniergues felt wanted. Shortly after arriving in Cuenca, he was with Ulloa out near the Tarqui baseline when there was an altercation with a young man. Ulloa was slightly wounded and he and Seniergues filed a formal complaint with a local magistrate, who arranged for the assailant to be arrested. Having discovered the young man's location, Seniergues and Ulloa rushed ahead of the officials, hauled the man from his hiding place and took him to Seniergues' lodgings, where the surgeon ordered his slave, Cujidón, to flog the man 200 times then rub pork fat into his open wounds. Jean Seniergues would not be the first medical practitioner to take the arbitration of pain to the wrong side of the line. Compassion and punishment had become parallel responses.

The expedition members continued to assemble in Cuenca. When Jussieu and Morainville showed up, they brought with them a botanical treasure chest of samples and descriptions of the cinchona tree. Jussieu's notes included the revelation that 'KinaKina' had been discovered by Malacatos people living in a hot, humid malarial valley some 60 miles south of the town of Loja. The suffering the Malacatos had endured from intermittent fevers had led them to experiment with plants from the local forest and, reported Jussieu, they found that the bark of cinchona was the 'almost unique remedy.' The Malacatos came to call the cure *ayac ca* ('bitter bark') or *yarachucchu carachucchu* (bark of the tree for fever chills). It was the Malacatos who had shared their arboreal remedy with a Jesuit missionary, who took the knowledge and the powder back to Spain, where it became the preferred treatment for the disease known as *mal'aria*.

Jorge Juan and Ulloa continued with their undercover investigation of the colonial administration. The geodesic survey had given the two lieutenants an intimate insight into everyday life across the Viceroyalty. They had slept in the homes of impoverished farmers and in the haciendas of the wealthy. They had hung out with muleteers and mayors, shepherds and presidents. And what they had seen was deeply unsettling. Among the humanitarian abhorrences they had researched was the *mita* system of forced labour, a corruption of an Inca practice that required communities to provide workforces for projects such as repair work to roads and bridges. Under Spanish colonial rule, *mita* had become a brutal tool that operated across the productive economy, from wool mills and mines to farms and sugar mills. Villages were required to provide fixed numbers of people on rotation, for a year at a time. Jorge Juan and Ulloa found that, on average, a worker on a hacienda was paid 18 pesos a year, from which was deducted an annual tribute payment of 8 pesos. The remaining 10 pesos had to cover the cost of a cloak of coarse cloth (6 reales a yard), leaving them 7 pesos 6 reales to pay for food and clothing for the family for the year, and for compulsory contributions to church festivals. Most workers would finish the year deeply in debt to the *hacendado*. There was no escape because 'the Indian's debt increases in proportion to the length of time he spends in the hacendado's service, and he remains a slave all his life, as do his sons after he dies'. In the *obrajes*, the wool mills, the two lieutenants witnessed overseers whipping workers with a yard of cowhide 'twisted together like a bass guitar string, and hardened'. The victim would be forced to remove his light trousers and lie face down on the ground and count the lashes, then kneel and kiss the hand of the overseer. The punishment was administered 'to all Indians – old men and young men, women and children'.

By August 1739, the Geodesic Mission to the Equator could claim to be the world's first multidisciplinary scientific expedition. True latitude and the shape of the Earth were about to be revealed for the first time. They had undertaken ground-breaking research on rubber and malaria. They had completed the first detailed survey of an Inca site. They had lifted the lid on colonial corruption and the oppression of native Americans. In a mine near Quito, Ulloa had come across a strange, silver-grey rock 'of such resistance, that, when struck on an anvil of steel, it is not easy to be separated'. The miners knew it as *platina*, a diminutive of the Spanish word for silver. The young lieutenant was the first European to describe platinum. Collectively, the mission had drawn new maps and taken thousands of measurements, recording altitudes, locations and temperatures. They were astronomers, surveyors, geographers and botanists, cartographers and physicists, medics and archaeologists. That August in Cuenca, they began to detect the light of success.

17. In the heart of the Andes, a 180° panorama from the Yaruquí baseline. In the centre are the twin peaks of Pichincha (see detail), scene of many mission dramas. The Pichincha signal, where La Condamine, Bouguer and Ulloa camped on a spike of rock for three snowbound weeks is marked at 'f'. The timber cross at 'e' was used by the mission as a navigational beacon in bad weather. Tanlagua, the peak that Ulloa slid down on his backside, is at '13'. The two ends of the baseline are marked at 'B' (Oyambaro) and 'A' (Caraburo). Cotopaxi is erupting at '2'. The observatory at Cochasquí, where Bouguer and La Condamine spent months crouched beneath the zenith sector, is just visible at 'k', on the far right below the volcano of Mojanda (16).

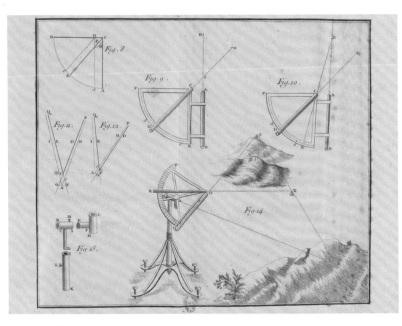

18. The GPS of the 18th century: the trusty quadrant was a place-fixer, used for measuring the angles between 'signals' placed at prominent locations. An attached telescope fitted with crosshairs allowed for accurate sighting. By tilting the quadrant on its iron stand, it could be used for measuring the heights of mountains.

19. Wild camping with a *marquise,* one of the large double-skin tents commissioned by La Condamine for use at high altitude. Two frock-coated surveyors attend to the quadrant while assistants climb an erupting volcano to replace a lost signal. The reality varied a little from this engraving in La Condamine's book. The tents got ripped and their poles broken, while the surveyors were unkempt and wrapped in Andean clothing.

TABLE du Calcul des Triangles

I. ORDRE & PLANS des TRIANGLES.	II. NOMS DES LIEUX où étoient posés les Signaux.	III. ANGLES DE POSITION observés.	IV. Equation pour la somme des 3 Angles	V. LONGUEUR des côtés opposés aux Angles observés.	VI. ANGLES de hauteur & de dépression apparente observés.
I.	Pamba-marca....	P. 38° 36′ 14″	−3″	CO. 6274,05 (Toises)	C.−5° 41′ 20″ calcul / O.−4. 30. 27. c.
	Carabourou........ *Terme Nord de la Base*	C. 77. 35. 40	−4	PO. 9821,00 (Base inclinée)	P.−3. 33. 6. b / O.−1. 6. 19. ab
	Oyambaro........ *Terme Sud de la Base*	O. 63. 48. 16	−3	PC. 9022,96	P.+4. 20. 29. a / C.−1. 11. 53. aa
		180. 0. 10	−10		
II.	Pamba-marca........	P. 69. 46. 37	0	OT. 15663,05	O.−4. 30. 27. c. / T.−1. 26. 20. c.
	Oyambaro........	O. 74. 10. 58	0	PT. 16060,29	P.+4. 20. 29. a. / T.+i. 18. 39. a.
	Tanlago1........	T. 36. 2. 25	0	PO. 9821,00	P.+1. 11. 13. d. / O.+1. 33. 48. d.
		180. 0. 0	0		
III.	Pamba-marca........	P. 38. 36. 34	−2	TΠ. 12690,77	T.−1. 26. 20. c. / Π.−0. 9. 53. b.
	Tanlagoa........	T. 89. 14. 10	−2	PΠ. 20335,92	P.+1. 11. 13. d. / Π.−2. 2. 56. b.
	Pitchincha........	Π. 52. 9. 22	−2	PT. 16060,29	P.−0. 28. 36. bd. / T.−3. 18. 8. c.
		180. 0. 6	−6		
IV.	Pamba-marca........	P. 39. 47. 3	0	ΠS. 13251,57	Π.−0. 9. 53. b. / S.−1. 21. 47. c.
	Pitchincha........	Π. 61. 6. 24	0	PS. 18131,07	P.−0. 28. 36. bd. / S.−3. 38. 56. cd.
	Schangailli........	S. 79. 6. 33	0	PΠ. 20335,92	P.+2. 4. 53¼ c. / Π.+3. 25. 47. c.
		180. 0. 0	0		
V.	Pitchincha........	Π. 58. 26. 10	−4	SC. 18097,10	S.−3. 38. 56. cd. / C.−0. 11. 16. cd.
	Schangailli........	S. 82. 57. 50	−4	ΠC. 21128,15	Π.+3. 25. 47. c. / C.+2. 24. 31¼ c.
	El Coraçon........	C. 38. 36. 12	−4	ΠS. 13251,57	Π.−0. 7. 35. c. / S.−2. 42. 10. c.
		180. 0. 12	−12		

20. Triangulation made simple: a sample page from La Condamine's 1751 book showing the angles observed between some of the early triangles. After measuring the length on the ground of the initial base line (C – O in top diagram), the following observations were angular.

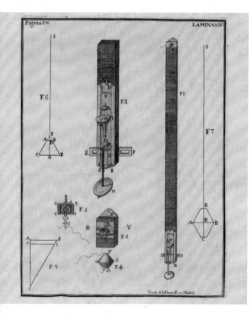

21. Pendulum problems: one of the more sensitive instruments shipped from France could be used for measuring Earth's gravity. It had to be mounted on a wall or a solid frame. Earthquakes and the friction of moving parts caused countless difficulties.

22. Fog-bow on Pambamarca: in this composite view, one of the early types of signal can be seen beside a *marquise* tent. The 'fog-bow' – or Brocken spectre – was first experienced by Bouguer when he saw his own image projected by low sunlight onto cloud, surrounded by rainbow-coloured 'coronets'.

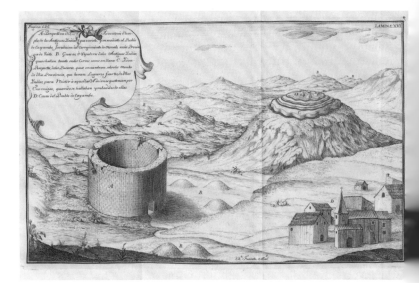

23. Lost worlds of Pambamarca, where the Spanish lieutenant Ulloa drew a 'temple' to the 'ancient Indians' and the concentric rings of what appeared to b a hillfort, where one of the mission's first signals was erected. In the foregrounc are the houses and church of Spanish Cayambe.

Euphorbiaceae
(Acalypheae)

Hevea brasiliensis Müll.Arg.

24. Rubber discovery: during his trek from the coast towards Quito, La Condamine stumbled upon a strange, elastic compound derived from the sap of a rainforest tree. Among the mission's several 'firsts' was the earliest European description of rubber.

25 & 26. The cure for malaria: Jussieu, Seniergues and La Condamine explored the forests around Loja where the cinchona tree thrived. With time, the powdered bark of cinchona became known as quinine. The inset drawing is from La Condamine's report, *On the Cinchona tree*.

27. The crater lake of Quilotoa: when La Condamine searched for a lost gold mine and a peak that might conclude the survey, he came upon a lake that – he was told – had once gushed fire and incinerated flocks of animals. He descended into the crater to investigate.

28. Chimborazo (6,310 metres) was thought to be the highest mountain in the world. Bouguer, La Condamine and Ulloa took instruments partway up the volcano to test Newton's law of gravitational attraction. The mission's latitude measurements confirmed that the summit of Chimborazo is the furthest point above the Earth's centre – and therefore the closest place to the stars.

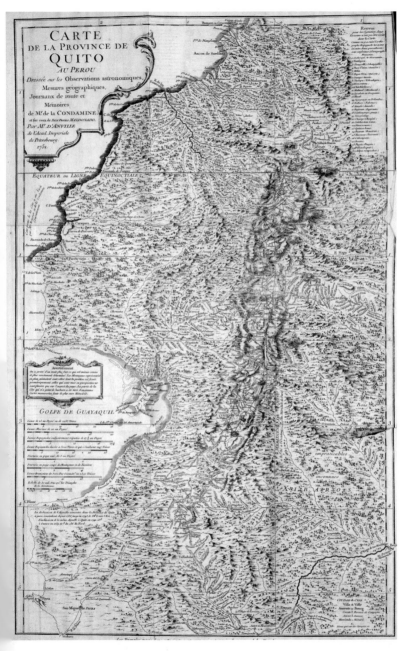

9. After four years of surveying, the chain of triangles extended for over three degrees of latitude at the equator, seen crossing the map, just north of Quito.

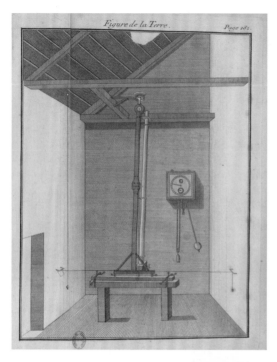

30. Instrument of torture: the zenith sector was so large and sensitive that it required its own observatory, in practice a building with a hole in the roof and with a pendulum clock on the wall. The observer had to lie on the floor beneath the eyepiece of the immensely long, precisely aligned telescope. More than three years were spent on the astronomy that located the ends of the chain of triangles.

31. The bulging globe in the book that Jorge Juan and Ulloa published about the expedition. The Geodesic Mission to the Equator proved that one degree of latitude was shorter at the equator than at the poles and the Earth was – as Newton had theorized – an oblate sphere.

12

Saturday, 29 August 1739 would be etched into the memories of the mission. It was fiesta time in Cuenca. For five days, the plaza in front of the church of San Sebastián ran with *chicha* and bull's blood. To accommodate the spectators, two decks of timber staging had been erected in a rectangle around the plaza, with gates in the corners for the bulls and processions. By the final afternoon of the fiesta, many of the spectators had achieved alcoholic saturation and were pumped for the climactic event: a dance of death which would end with a thin, pointed *verduguillo* being plunged into the neck of a slathering bull. Instead of an afternoon on the edge of their seats, they got an interactive riot. And the *verduguillo* ended up in a Frenchman's guts.

It was a tragedy that had been developing for several weeks. The repeated excursions to the bleak heights of the cordilleras carrying strange devices had become a source of local curiosity. Few believed that the visitors were not prospecting for gold or silver, a suspicion reinforced by their bleached and weathered clothing. And there were dalliances with local women. Sixty years later, when the German naturalist Alexander von Humboldt passed through town, he would be told that La Condamine had fathered two daughters in Cuenca.

And then there was the rumoured relationship between the French surgeon and the jilted lover. Jean Seniergues was said to be romantically involved with Manuela de Quesada, the daughter of Francisco de Quesada, who had been treated for malaria by the surgeon. Manuela had been dumped by her fiancé, Diego de Léon y Román, the town's deputy attorney-general. Francisco and Manuela expected the dastardly Léon to compensate the family for its loss of dignity by honouring them with a payment. With a provocative lack of foresight, Seniergues then triangulated the row by intervening and demanding that the deputy attorney-general hand over the money. Léon responded by sending his slave to the Quesadas' house with instructions to slap Manuela and to challenge 'her Frenchman' to 'remove the blow'. Seniergues – standing by his story that he had come to the Quesada household to cure Francisco rather than to court Manuela – beat the slave with a stick and sent her back to Léon with a demand for 'satisfaction'. It was a bad day to be a slave, but even worse for Seniergues. Challenging Cuenca's well-connected deputy attorney-general to a duel was not in the interests of Franco-Spanish diplomacy, nor likely to further the mission's scientific endeavours.

The five-day fiesta was already underway when, on the afternoon of the 26th, the French surgeon spotted Léon in Cuenca's main plaza and challenged him to draw his sword. Léon whipped out a pistol. In an encounter characterized by alcohol rather than fighting skills, Léon's flintlock misfired and Seniergues tripped into the gutter. As the surgeon was escorted to his lodging, amused bystanders heard him threatening to kill Léon, or to 'cut off his ears'. In an attempt to defuse a dispute that had every likelihood of leading to blood in the streets, a Jesuit priest invited the two factions to meet at his lodgings on the 28th. Léon failed to appear. Anger, retribution and *chicha* were perfectly pitched for an unhealthy showdown.

For the final day of the fiesta, the members of the mission had been invited to enjoy the bullfight. Fatally, they took their seats in different sections of the stands. While the crowd waited for festivities to start, Seniergues lurched into the plaza and headed towards the section of stand occupied by Manuela and her father. The crowd quickly made up its mind: Francisco de Quesada was being far too friendly with an arrogant Frenchman who had challenged one of their own to a duel and was behaving improperly with another. As Manuela's father was dragged away by one of Léon's enraged friends, Seniergues leapt to the rescue, pistol in one hand and cutlass in the other, shouting at Léon: 'I'll kill the rogue and his whole family!' And to make sure the entire crowd understood that he was serious, he told his slave Cujidón to 'Kill them all!'

With imperfect timing, the fiesta's master of ceremonies entered the plaza on a magnificently brocaded horse. Nicolás de Neyra y Perez de Villamar was not a man to cross. He was both a captain in the militia and well known to Cuenca's elite. On seeing one of the Frenchmen making an embarrassment of himself, Neyra rode across to the stand where Godin was sitting with Jorge Juan and Ulloa and asked them to restrain Seniergues. Neyra then crossed to Seniergues and attempted to pacify the surgeon. But it was too late. Seething with fury, Seniergues told Neyra that he would kill him too. Then he overturned the table and vaulted into the plaza. Publicly humiliated, Neyra announced to the crowd that the bullfight was cancelled. Then he departed in search of the town's *alcalde*, whose mayoral authority was required to arrest and imprison Seniergues.

Denied their bullfight, the crowd turned on *la compagnie françoise*. The plaza erupted into an arena of anger. Jorge Juan and Godin remonstrated with Seniergues. Heads turned towards a new commotion as a troop of at least a hundred men armed with swords, pikes and lances poured through the northern gate, led by Neyra and the *alcalde*, Sebastián Serrano de Mora y Morillo de Montalban. Ordered to hand over his weapons, Seniergues raised his cutlass and cocked pistol. Pointing the double barrels at Serrano, he pulled one trigger and then the other. Both barrels misfired. Undeterred, the French surgeon brought his cutlass down in a curving swipe. The blow was deflected by one of Serrano's entourage and then Neyra pierced Seniergues' hand with the point of his sword. As the pistol fell to the ground and Serrano called for Seniergues to be seized, the crowd joined in, shouting for the 'foreigners' to be killed. Stone setts were wrenched from the plaza. Seniergues retreated before an onslaught of missiles and advancing steel. A blow to his right arm knocked the cutlass from his grip. Defenceless and bleeding, he made for the gate at the far corner of the plaza, but before he could haul himself clear, Neyra snatched a short-bladed *verduguillo* and deftly slipped it between the Frenchman's ribs. Pursued by the armed mob, the surgeon staggered into the courtyard of a nearby house, where he collapsed.

In the street, La Condamine and Bouguer ran towards the house, where Seniergues lay unconscious, but their way was blocked by a crowd shouting, 'Kill the French foreigners!' Bouguer was struck by a rock. A sword-thrust pierced him in the back. A quick-thinking priest yanked the Academician into a house and barred the door. Dispersed members of the mission legged it through the streets. 'There was not one of us,' remembered La Condamine, 'who did not run the risk of life, and the Spanish officers, our travelling companions, were not exempt from the same peril.'

In the chaos of dust, drumming feet and yelling, the Europeans were hustled into doorways as the tumult raged past. La Condamine made it back to his lodgings, where his barred door was besieged by the mob.

In the house where Seniergues lay bleeding, Serrano managed to stem the rush towards the courtyard. The surgeon was moved to a room with a bed. When Joseph de Jussieu managed to gain admittance, the young doctor examined Seniergues. The only wound of serious concern was the neat, deep *verduguillo* cut.

In four years of adventuring, the mission had been confronted by a variety of mishaps and disasters. The loss of Couplet-Viguier had reduced the original team to eleven. Now Bouguer and Seniergues had been felled by blades. Bouguer's wound was unlikely to be fatal. But the cut in Seniergues' side was more worrying. He was bleeding internally.

Two days after the fracas in the plaza, Jussieu poured his feelings into a letter home to his brothers: 'We French were almost overwhelmed by the uprising,' he wrote, 'and barely escaped being killed. Seniergues alone has paid for all of us. His state allows me not a moment's rest, and I am at once pharmacist, surgeon and doctor.'

Among the Frenchmen, the only one to have given up hope was Seniergues himself: 'It's in a very bad place,' groaned the surgeon. 'I am lost!' He could feel the cold tendrils of mortality invading his body. The wound was infected.

Seniergues dictated his last will and testament, appointing as executors his friends Jussieu and La Condamine. He asked to be interred in the city's Iglesia Matriz cathedral. He listed debts to pay and charities to support. He left his two slaves to La Condamine. He tidied his time on Earth and died in La Condamine's bed.

The murder of Seniergues was an act of geodesic sabotage. As the fractious Academicians were about to attempt their critical metamorphosis from mountaineering geometers to observatory astronomers, the entire mission was robbed of momentum.

For Seniergues, Peru had been the back door to entitlement. He had sailed from Rochefort dreaming of gold mines but had fallen in a dusty bullring. Had the mission been a better-led unit, Seniergues would have been given less latitude. But the status and the cash he had accumulated through private medicine – and perhaps the stress of continual displacement – had infected his personality. His role in the brutal flogging of the man at Tarqui had been that of a psychopath. He had become addicted to hedonic reward: aggression made him feel powerful, dominant. The company of lieutenants cannot have helped. Jorge Juan and Ulloa were warriors who had seen off assailants in Quito's plaza and been excused for the death of the president's secretary. Seniergues was not in their league. He had misread Cuenca's crowd and failed to land a single blow on his assailants. In the theatre of aggression, he had proved himself a tragic actor rather than a skilled practitioner.

Bouguer, whose greatest worry before the expedition had been seasickness, recovered quickly from his sword wound. For La Condamine and Jussieu, the healing was more complicated. As named executors of Seniergues' estate, they both had administrative duties to discharge. Jussieu, who was not robust at the best of times, carried the additional burden of having been the bedside medic during the four days that it took for Seniergues to die. La Condamine regarded the death of his friend as a crime for which culprits had to be punished. He embarked upon a legal campaign that would soon add to the mission's accumulated instabilities. As executor and legal dilettante, La Condamine was sucked into a time-eating vortex. The mission had lost more than its surgeon. The man who had so often been the engine of recovery became distracted by legal wrangles arising from the *tumulte* in Cuenca.

For them all, Cuenca had flipped from safe haven to health hazard. The confrontation in the plaza had mutated into a full-scale riot. Provisional control of the town had been restored by armed patrols and by an edict banning gatherings, but the criminal complaint filed by Bouguer and La Condamine had named three of Cuenca's better-connected citizens: the mayor, the deputy attorney-general and Neyra, the militia captain. The streets of Cuenca remained dangerous for anyone associated with *la compagnie françoise*.

There had been many interruptions to the survey, but resumption had always been relatively straightforward because the process of measuring triangles had been repetitive and scientifically uncomplicated. Now, the mission had to restore its sense of direction and focus on the dark science of astronomy. And to do that, they needed to regain their scientific composure and return to their most troublesome instrument, the zenith sector.

Compared to the mountain trials of the previous two years, the observations required to complete the mission were – at first glance – little more than an astronomical postscript. All they needed to do was to take the star sightings necessary to fix the precise latitude at the southern and northern end of the chain of triangles. The difference between the two latitudes would equate to the angular extent of the chain in degrees. By dividing the total measured length of the chain in *toises* by the astronomical extent in degrees, they could find the figure for one degree of latitude. The three Academicians could be forgiven for thinking that the hard work was behind them. But as ever, they created an additional tier of difficulty: instead of fixing latitude from a single location at the southern end of the chain of triangles, the twin baselines meant they would have to establish two observatories. The doubling up promised greater accuracy, but it guaranteed more delays. There would be two teams of astronomers. And they would need two zenith sectors.

The mission's instrument-maker, Théodore Hugo, found himself on the front line of operations. Some time earlier, Godin had instructed Hugo to build an entirely new eighteen-foot sector. For a clockmaker, this was a considerable challenge. Hugo had to cast, mill and turn components in iron and brass, then assemble a huge precision instrument from an array of lenses, suspension brackets, tubes, bars, screws and the all-important pivot. Godin was certain that his new instrument would hasten his observations. The battered old twelve-foot Graham sector would be used by Bouguer and La Condamine. Hugo was tasked to repair the instrument and to improve its means of suspension.

Godin wasted no time. The southern baseline at Baños was connected to the main chain of triangles by signals that included the tower of Iglesia Matriz cathedral, so Godin was able to set up his observatory in

Cuenca. His new assistants were the mission's trained killers. With Jorge Juan and Ulloa as a two-man close-protection squad, the ongoing antipathy towards *la compagnie françoise* was less likely to trigger further losses. In a house close to the centre of town, the giant new sector was carefully set up and aimed at the stars.

While Godin could observe the stars from the convenience and comfort of a town, the observatory on the Tarqui baseline was bleak and remote. Observations were delayed by the repairs Hugo needed to make to the twelve-foot Graham sector, and it wasn't until the beginning of October that Bouguer, La Condamine and Verguin took the rocky road south towards the foothills of the cordillera. The only available building they could find close to the Tarqui baseline was an isolated part-built chapel, which they began converting into an observatory. The repaired Graham sector was not installed and ready until mid-October. Conditions at the chapel were miserable. Most supplies had to come from Cuenca, 17 miles to the north, along a road that forded rivers five times. As ever, the locals were curious and the astronomers got used to working with an audience.

Night after night, week after week, the teams at the Cuenca and Tarqui observatories peered skyward, waiting for breaks in the cloud cover and filling their notebooks with columns of figures. They were focussed on one of the brightest stars in the sky, Epsilon Orion. On clear nights, it glittered like a jewel and the great arm of the sector could be carefully moved so that the cross-hairs in the telescope intersected with the Epsilon's distant twinkle. Mostly, the weather was cloudy. And there were endless problems with the operation of the sectors and with discrepancies in the sightings.

At the Tarqui observatory, readings from the graduated scale on the sector's limb varied from night to night by 8 to 10 seconds of arc, when the same star was being observed. The anomaly stood in the way of an accurate latitudinal fix for the Tarqui baseline. In this lonely billet, La Condamine slipped into misery. He recorded the interminable exercise as a 'series of sad & painful observations'.

The only moment of respite from the exhausting nocturnal vigils was provided by Tarqui's annual fiesta, which was run with gusto by the local community. Fancy-dress horseraces with spear-wielding riders were followed by a pantomime. It took a while for the astronomers to realize that the performers were re-enacting a geodesic survey:

> I had seen them several times watching us attentively, while we took the heights of the sun to set our clocks. It must have been an impenetrable mystery for them, that an observer kneeling at the foot of a quadrant, his head thrown back, in an awkward attitude, holding in one hand a smoked glass, handling the screw of the foot of the instrument, bringing his eye alternately to the telescope & the division, to examine the plumb line, running from time to time looking at the minute & the second on a pendulum, writing some figures on a paper, & resuming its first situation.

For weeks, the people of Tarqui had been watching

the bizarre antics of the French astronomers. When the young performers produced giant quadrants of wood and paper and re-enacted the rituals of geodesy, La Condamine relaxed at last and laughed at mimicry so accurate 'that it was impossible for us not to recognise ourselves'. Later, he related that he had seen 'nothing more pleasant during the ten years of the journey'.

Up the road in Cuenca, Godin was also baffled. His new eighteen-foot zenith sector was providing anomalous results. On the all too infrequent nights that allowed observations, he was not observing a constant height for Epsilon Orion. Endless adjustments to the sector failed to eradicate the variations. The astronomical challenges were not helped by tension in the streets. With La Condamine pursuing the perpetrators of the riot through the courts, there were many in town with a simmering antipathy towards *la compagnie françoise*. This erupted one night when Godin, Jorge Juan and Ulloa were forced to undertake a critical measurement that took them outside the observatory. To maintain absolute accuracy, they needed to pace the distance of their observatory from the tower of Iglesia Matriz cathedral. Without that measurement, they could not connect the new astronomical sightings at the observatory to the chain of triangles. Sneaking out under the cover of darkness, the trio were spotted by some women, who rallied the populace and drove the astronomers back to their quarters with sticks and stones. In December, after three months clamped to his zenith sector, Godin closed the observatory door and left for Quito.

At Tarqui, Bouguer, La Condamine and Verguin continued their vigil. They had started later than Godin, and it was not until early January that Bouguer declared that he was ready to leave the observatory and travel back to Quito, where they would establish a new observatory. La Condamine could sense that the end was in sight. The forthcoming series of astronomical observations at the northern end of the meridian would determine the latitudinal span of their arc. After that, it was just maths. Then France.

Not for the first time, La Condamine did his own thing. He stayed on for a couple of weeks at Tarqui with Verguin and Graham's old zenith sector, trying to resolve the discrepancies in the observations and unsure whether they were due to faults in the instrument or some kind of astronomical aberration. He also experimented with the effect of air resistance on pendulums, by redeploying the tiny gold ball he had asked Hugo to cast three years earlier. For four hours, the glinting two-ounce sphere swung to and fro in Tarqui's mountain air, nearly an hour and a half longer, he noted, than had been possible with a larger copper ball of the same weight. They left Tarqui on 16 January, with the zenith sector dismantled and packed for the 270-mile journey to Quito. To avoid damage, it would be carried by porters all the way.

La Condamine saw it as a last opportunity to appreciate a passage along the Great Road of the Incas linking Cuenca and Quito and to bid a long goodbye to a land he had known for four extraordinary years. Over three weeks, he ambled northward with Verguin, dropping in on old friends and tying up loose ends. They paused for a few days in Cuenca, where La Condamine conducted declination experiments with a new magnet made by Hugo. He gathered affidavits from the town's religious orders and curates; men who had been sympathetic to the mission during and after the Seniergues affair. The testaments were a necessary cleansing of the mission's record. The two men made a side trip to Cuenca's hot springs, collected samples of the water and then continued north to Riobamba, where they were invited by the Maldonados to a wedding in a country house.

La Condamine's baggage had been delayed on the road and he had to attend the wedding celebrations at San Andrés in his grimy travelling clothes. He remembered it as *'la plus magnifique & la plus brillante'* feast he had experienced during his entire time in South America. For three heady days he enjoyed himself on what he claimed was his longest holiday of the entire expedition. When the baggage caught up, La Condamine retrieved his quadrant and rode with Verguin to the beautiful shores of Lake Colta, where the two men used the mirror-like surface to conduct some experiments on refraction. A more idyllic spot would be impossible to imagine. After the miseries of the past months, it was all too good to last. And, of course, it didn't.

From the 'enchanted island' of San Andrés, La Condamine and Verguin rode north up the Great Road. When they reached Quito, on 7 February 1740, they found that Bouguer had arrived several days earlier. Not a moment too soon, the dismantled zenith sector from Tarqui was carried into town by a string of porters. Bouguer put the indispensable Hugo to work, stiffening the sector's frame in preparation for its final deployment.

On the 11th, Bouguer rode north out of town with Verguin, bound for the slopes of Mojanda, where they planned to set up their final observatory. It was a district Verguin knew well. He had undertaken a reconnaissance of Mojanda's stepped slopes back in June 1736, when he had been looking with poor Couplet for a level plain that could be used as the northern baseline. Four years later, young Couplet was dead and he was back on Mojanda with Bouguer, the old man of the expedition. Beyond the sloping plateau of Malchinguí, they plodded up the switchbacks to the crest of the rib bearing the scattered homesteads of Cochasquí and the strange overgrown pyramids built by an ancient culture to observe its cosmos. La Condamine – who had been delayed in Quito by criminal proceedings related to *l'affaire de Cuenca* – reached Cochasquí a few days later and pronounced its situation 'very favourable: we could clearly see our first base, at its two extremes, as well as all the signals of the surroundings'.

For ten weeks, they took turns to crouch beneath the eyepiece of the zenith sector, meticulously logging the zenith and the timing of the stars that would allow them to compute the exact latitude of the observatory. As ever, they were slaves to the rain, fog and cloud, but by the end of April Bouguer declared that he was content with the observations.

All they had to do now was to connect the observatory to the chain of triangles with a couple of quadrant measurements. They split: Bouguer took his quadrant across to Tanlagua and measured the angle between Oyambaro and Cochasquí. La Condamine rode across to Oyambaro and measured the angle between Cochasquí and Tanlagua.

The adventure was over. After four years of a wandering life,' mused La Condamine, 'two of which were spent in the mountains, I returned to Quito on May 1, 1740, with the intention of drawing there at leisure the consequences of all our measurements, & to conclude the value of the degree of the meridian, which was the goal of so many operations.'

Back in Quito, Bouguer worked the numbers. Epsilon Orion had been observed from Cochasquí to be 1° 26′ 38″ south of the zenith. From the Tarqui observatory, the angle had been 1° 40′ 35″ north. So the length of the arc was 3° 6′ 43″. There were a few more figures to compute, such as the change in declination of Epsilon Orion since January and the effect of parallax. But the astronomy was concluded.

On 6 May 1740, Bouguer signed a document before a notary confirming that he regarded 'the object of our mission as entirely fulfilled'. The document was countersigned by La Condamine and by Verguin. But not by Godin.

Three years later, they were still in South America. Still staring at the sky.

13

Hunched over the workbench in his Quito *atelier*, Théodore Hugo faced the challenge of his life. Louis Godin, *le patron*, had ordered the clockmaker to construct another new zenith sector. A huge zenith sector. A zenith sector with a twenty-foot radius. It would have been a demanding commission for a professional instrument-maker in London, but in Quito, where the metal alloys and parts for astronomical instruments were rarer than gold dust, it would take time. He needed precision-ground lenses, sheets of brass and copper, castings in iron. There would be soldering, and threads to cut with his taps and dies. And the whole, gigantic, delicate instrument would have to be designed in such a way that it could be dismantled for Andean transport then reassembled in a distant observatory.

Louis Godin wanted to start again. During his sojourn in the Cuenca observatory, he had become aware of what appeared to be variations in the height of Epsilon Orion. Repeated adjustments of the eighteen-foot zenith sector had failed to settle the wandering star. To remove the possibility that the variations might be caused by flaws in the zenith sector, Godin wanted to take a larger, more accurate instrument back to Cuenca and try a second season of observations.

All thoughts of an imminent return to France were lost in Peruvian mists. La Condamine claimed that both he and Bouguer had also observed 'odd, and sometimes very noticeable, changes from day to day in the height of the stars near the zenith'. Both men wondered whether the variations were caused by tiny movements transmitted to the instrument from the diurnal contraction and expansion through cold, heat and humidity of the mud-brick wall to which the zenith sector was attached. For Bouguer, the variations were not enough to keep him from a return to France. But he could not leave Peru if Godin intended to resume observations. Yet again, the three Academicians went solo.

While Godin waited in Quito for his new zenith sector to be constructed, Bouguer took a trek to the coast. From the start, they had agreed that the chain of triangles would need to be mathematically levelled to a horizontal plane at sea level in order to iron out distortions created by irregularities in the heights of the various signals. To do this, they needed to know the exact height above sea level of one of the established signals. Two years earlier, La Condamine had tried to connect the chain of triangles to sea level with his inconclusive excursion to Quilotoa, from whose heights he had hoped to see the ocean. Bouguer's latest plan was to ride Maldonado's new road down the Esmeraldas valley to the coast, taking height measurements with a barometer and then observing with a quadrant the summit signal of Pichincha. He would take his servant, Grangier, as his assistant. Like so many of the mission's excursions, it was simpler in conception than in execution.

One month after setting out from Quito, Bouguer sent a message to La Condamine saying that they were on a small island in the estuary of the Esmeraldas and that jaguars had raided the camp and eaten the food. The humidity was oppressive and Pichincha had appeared only once, for three minutes, which had proved insufficient time to grab the necessary observation.

Left behind in Quito, the gung-ho war vet had lost his mojo. Later, he was accused of shirking his duties by failing to accompany Bouguer to the coast, a slight he countered by claiming that he was kept in Quito by 'the sad occupation' of concluding his astronomical calculation and by 'all the lawsuits'. The maths was far worse than mountaineering:

> I felt terrified at the sight of the long calculations I was going to undertake . . . I had an extreme repugnance for a job which little habit makes painful & repulsive when one is not familiar with it, while for the practiced calculator, it is only a gentle and peaceful occupation. It can even cease to be boring to him, by the promptness with which he finds the results he seeks . . . What advantage to reach the end, the one who knows the shortest way, and who is sure never to take a wrong step! I admit that what would perhaps have been only the work of a few weeks for another, cost me several months.

He had been good with compass and quadrant. He had mastered the pendulum and the zenith sector. But sums were a pain in *le cul*. Godin was not sharing his results and – in the absence of Bouguer – La Condamine wanted to compute his own answer to the latitude question.

With the wearisome mathematics and the interminable court case against Serrano, Léon and Neyra, La Condamine had also become *chef d'affaires* of a pyramid scheme. It had been conceived in the Louvre before *Portefaix* sailed from Rochefort and it was part of the Academy's ambition to embed Gallic geodesy in South American soil and to acclaim the scientific credentials of King Louis XV. Before the mission left France, one of the Academy's sister institutions – the Académie des Inscriptions et Belles-Lettres – had furnished Godin with wording to be inscribed on a monument 'to mark the Centre of the Operations . . . at the Equator'. Over time, La Condamine had become the mission's monuments-man. He had started in 1736, with the inscription he had cut into the coastal boulder at Palmar on the equator. And for over a year, he had been making arrangements for a far grander, double monument to be raised on the baseline that had launched the chain of triangles four years earlier. While stuck down at Tarqui, he had sourced some stone and had it transported up the Great Road to Quito. And while he had been on his recent trip to the signal at Oyambaro, he had stopped at El Quinche to see his old friend José Antonio Maldonado, who had provided the French Academician with the materials and workers to raise 'two lasting monuments': a pair of gigantic pyramids, one to be erected at each end of the Yaruquí baseline on top of the two buried millstones. The mission's artist, Morainville, agreed to oversee the project. The Academy's inscription was to be cut into the two stone tablets quarried near Tarqui. La Condamine became obsessed with the project, writing in May 1740 that a 'large part of the rest of the year was spent in trips that [I] made to Quinche, to the baseline and to the surroundings, to give the necessary orders for the work.' The pyramids were a time-consuming – and ultimately disastrous – distraction.

Saturday, 27 August 1740 dawned in Quito with an ominous rumble. Walls trembled, dogs barked. It was a big one. Charles-Marie de la Condamine fell out of bed and braced himself for the aftershocks of a *tremblement de terre*. Things were looking ragged. Godin was away in Cuenca with Jorge Juan, Ulloa and the new twenty-foot zenith sector. Bouguer had not returned from the Esmeraldas expedition. Three weeks earlier, war had reached Quito when several hundred mules loaded with gold and silver had filed into the city, along with dust-caked officials from Lima and various agents of Spain and Peru. La Condamine wrote of 'lively alarms'. Portobelo had been captured by the British. Cartagena de Indias had been attacked. San Lorenzo fort on the Chagres had been destroyed. Separated from the coast by Andean barricades, Quito became the Viceroyalty bank vault.

Amid the chaos of earthquake and war, La Condamine consoled himself that the chatterers, troublemakers and *nouvellistes* of Quito were at last diverted from interfering with science. It had never been possible to convince the people of Peru that the mission's antics with instruments were devoted to knowledge rather than greed. The general mayhem and astonishing transformation of Quito from provincial backwater to the 'depositary of most of the wealth of the new world', put an immediate end to the constant pestering about the mission's purpose. Geodesy had been overtaken by geopolitics.

The evening of the earthquake, Bouguer finally returned from the coast. He had been gone for more than three months. The two Academicians caught up with each other's news. La Condamine told Bouguer that he had almost completed his calculations. Bouguer said that he had 'a few angles' to measure in order to connect his coastal observations with the chain of triangles. This would entail a return to the signal on Papa-urco, just south of Cotopaxi. Both men were 'seriously thinking of [our] departure for France'.

Through September 1740, while Bouguer headed off to the heights of Papa-urco to take his angles, La Condamine busied himself with the various items of outstanding business that he wanted to resolve before leaving for France and home. He was keen to extend the speed-of-sound experiment by measuring the time taken for a detonation to travel over a greater distance than they had achieved back in 1737 and 1738. With permission from Quito's governor, he had a cannon taken to Guápulo, just outside the city, where he pitched his tent in readiness for Bouguer's return. He went to El Quinche on pyramid business. He took some observations on magnetic declination and he persevered with the maths. By the 20th, he had come up with a provisional figure for the length of one degree of latitude at the equator. Verguin, too, had done the sums. They shared homework and La Condamine was pleased to find a difference of 'only a few seconds'. They were still waiting for Bouguer to return with the angles that would allow the chain of triangles to be reduced to a horizontal plane, but could satisfy themselves that they were but a few hours from completing the task that had occupied them for so many years. Godin could go on tinkering with his zenith sector, but Bouguer and La Condamine would have all the numbers necessary to make the final calculation.

With departure for Europe now imminent, La Condamine packed another consignment of treasures for shipping back to France. Into the trunk went colourful clay antique vases, similar to the ones he had sent from Lima three years earlier, some with feet of silver and others with patterns made with burning coal. There was alabaster from Cuenca and various mortars and stone axes once wielded by people La Condamine called the *'anciens Indiens'*. Into the case went the fruits of three years of Andean travel: encrusted stones from the stream on Tanlagua, crystals, marcasite, two pieces of fossilized wood, a stuffed coral snake with 'rings the colour of fire & black', and a 'little crocodile from the river of Guayaquil'. La Condamine addressed the trunk to M. du Fay at the Jardin du Roi, assured that this diverse collection of 'curiosities of all kinds and monuments to the industry of the ancient Indians' would impress and enthuse his fellow Academicians in the Louvre.

La Condamine entrusted his trunk to Godin's young cousin, Jean-Baptiste Godin des Odonais, who had been so indispensable during the two years of triangulation. 'His duties related to the object of our mission,' noted La Condamine, 'had ceased.' Young Jean-Baptiste was free, but he was also marooned in Peru with no funds to get back to France. With his signal-fixing role concluded, Godin des Odonais had decided to take a commercial trip to Cartagena de Indias, where he planned to buy textiles to sell in Quito. He left on 3 October for the long ride north, taking La Condamine's trunk of treasures with him. The same day, Louis Godin reappeared in Quito with Jorge Juan and Ulloa. They brought bad news.

Two months in the Cuenca observatory working with the new twenty-foot zenith sector had been abruptly halted by a command from Viceroy Villagarcía that the two Spanish officers travel immediately to Lima, where their military skills were urgently needed to help defend the city. The British had reached the Pacific. It was the beginning of a fresh episodic crisis for the mission. Godin could not continue working in Cuenca without the protection of Jorge Juan and Ulloa. Their summons to war put a stop to astronomy.

Five days after the two Spanish officers left for Lima, La Condamine persuaded Bouguer to help out with the sound experiment. Verguin was on Guápulo, manning the cannon loaded with a nine-pound ball. Bouguer and La Condamine were watching through a telescope at a range of 10,540 *toises* (about 12 miles). They repeated the experiment three times and measured the speed of sound at 1,745 *toises* per second. The detonations took more than sixty seconds to reach La Condamine and Bouguer, arriving as faint, defiant thuds.

Bouguer's secret observatory was a body-blow for La Condamine. He had known nothing about it. On 2 November 1740, during an exchange about the incomplete observations, Bouguer led his fellow Academician to a 'remote location' at the edge of the city. Through a locked door, La Condamine was astonished to see a fully functioning astronomical observatory 'and the sector all set up.' For six weeks, Bouguer had been repeating the observations taken earlier in the year at Cochasquí. It would have been an interesting moment to be a bystander. Did Charles-Marie give him an earful? Wrap a ruler around his skull? In his memoir, La Condamine claims only to have asked his old *compatriote* for 'news of the observations' and requested that he should be allowed 'to take part'.

Bouguer still had observations to take on Papa-urco, so he left La Condamine the keys to the secret observatory, telling him that he would have to hand them back in six weeks, when he returned from the mountain. Grateful to be a working astronomer again, La Condamine moved into the observatory and returned to the rituals demanded of the zenith sector: nocturnal vigils in wait of clear skies; the excruciating posture beneath the low-level eyepiece of the telescope; the continuous checking of Bouguer's pendulum clock, which kept losing time, due – La Condamine concluded – to the humidity caused by the frequent rains. For occasional company and assistance, he had his slave, a man whose identity has been smothered by the prejudices of the age. Possibly he was one of the slaves who had worked for Seniergues. After a few days, La Condamine had his bed transported through the streets of Quito. He became an observatory recluse. The weather was awful.

This was an uncomfortable, despairing time for the once-ebullient explorer of river, peak and *páramo*. He was exhausted and affected by recurring fevers. One dark night, while he was alone in the observatory waiting for Epsilon Orion to climb to its zenith, the door to the street creaked open to admit the black shadow of a man holding a lantern. The door had been locked. Behind the figure with the flickering lantern were seven or eight other men with raised swords and pistols. Sure that he was about to be dispatched by the same kind of lynch mob that had murdered Seniergues, La Condamine froze beneath the eyepiece of the zenith sector. But the visitors did not fall with blade and barrel upon the helpless astronomer. After a confused exchange, the leading man revealed himself as the commander of a night patrol. He hadn't known that the house was occupied and had forced the lock. Later, when La Condamine was able to see the funny side of the incident, he related that the commander had departed 'rather ill-paid for his curiosity'.

From 9 until 23 November, the night rains were so continuous that La Condamine failed to catch a single glimpse of Epsilon Orion. Then, before the end of the month, Bouguer showed up from Papa-urco and announced that he needed his pendulum clock for some observations. The clock was dismounted from the wall and La Condamine was forced to install his own pendulum clock, losing yet more time while he carefully set its time by taking noon-day sun shots. Having hung and set his clock, he was back at the zenith sector by the end of November. A few nights later, he was lying unconscious on the floor of the darkened observatory.

Later, La Condamine self-diagnosed his condition as 'compression of the carotid arteries, caused by extension of the neck'. The to-and-fro shuffle from a prone posture on the floor beneath the telescope to an upright stance beneath the pendulum clock had caused him to black out. The zenith sector was trying to kill him: 'Fortunately . . . my Negro, who happened to be present, came to my aid: he told me that I had got up, and that I had fallen a second time.' As if unconsciousness was not hindrance enough, La Condamine was having trouble focussing the telescope.

They were all struggling. A few blocks away, Jussieu was on his deathbed. At the beginning of December, the doctor had been struck down by *une fièvre maligne*. Nothing Jussieu tried would quell the fever. Already, he had successfully treated many in Quito for an epidemic then raging through town. His own condition worsened and he was forced to 'put his affairs and his conscience in order'. Unlike poor Couplet, Jussieu managed to sweat off the fever and pull through.

Locked away at the edge of town, La Condamine was almost out of time. Bouguer wanted him out of the observatory by 16 December. As the deadline rushed towards him, and with no satisfactory observations of Epsilon Orion in his notebook, La Condamine pleaded for an extension. He managed to hang on until the end of December, without having been able to identify the cause of the apparent variations in height of Epsilon Orion. 'I could not,' wrote La Condamine, 'draw any conclusions from my work.' Another three months had evaporated.

Then, as the year turned to 1741, Louis Godin launched an epistolary meteorite towards his fellow Academicians. He wanted to scrap the astronomical observations from the southern and northern ends of the chain of triangles. The clock would turn back to 1739 and they would start over again.

This time, they would conduct an astronomical ambush and shoot Epsilon Orion from both ends of the triangulation chain simultaneously. If they could succeed in measuring the star's zenith on the same day from both the northern and southern ends of the meridian arc, they would remove from the results any variation that might have been caused by the movements of the star itself. Bouguer would take Morainville and his twelve-foot zenith sector south to Tarqui. Godin would take Hugo, Verguin and his new twenty-foot sector beyond the northern extremity of the chain of triangles, where he planned to set up a new observatory, on a hacienda near a town called Mira. Later, they would have to connect Mira to the chain of triangles by additional surveying.

La Condamine's role in this new operation was secondary: he would stay in Quito with his fifteen-foot fixed telescope as a check against the two sectors. Trapped in the never-ending expedition, La Condamine wrote home, explaining that the three Academicians would be 'detained in the country for a few months'. He assumed – rashly – that it would be the last letter he wrote from Quito to the Academy before being reunited with them in person. On 9 February, Bouguer left Quito on what he fervently hoped would be his final excursion along the Great Road to Tarqui.

Because Godin's new observatory at Mira was so close to Quito, he was able to delay his departure. It wasn't until 2 March that Hugo and Verguin rode out of the city with a column of porters carrying Godin's dismantled twenty-foot sector. The assumption that it would take just a few days to reassemble, adjust and calibrate the instrument proved optimistic. Two weeks after leaving, Verguin and Hugo had not managed to trace the all-important meridian on which the sector had to be aligned.

Back in Quito, Godin was reluctant to relinquish the comforts of the city. It was at this point that La Condamine executed a subtle *coup d'état*. The pretext was the raising of the two commemorative pyramids on the Yaruquí baseline. He persuaded Godin to accompany him to the palace of the *audiencia*. There, the president and his advisers learned from the two French Academicians that henceforth La Condamine would 'remain in charge, on behalf of the other Academicians, of all that concerned the construction of the pyramids'. La Condamine was able to produce a power of attorney, signed by Bouguer. In this revised hierarchy of leadership, La Condamine had taken charge formally of the French legacy in Peru.

By April, the two teams were ready to begin their observations. At the southern end of the arc, Bouguer had a difficult start. As he knew well, the Tarqui observatory was a tough posting, exposed to mountain weather and too far from Cuenca for comfort. To avoid missing a passing star, Bouguer had to drag himself out of bed several times each night, stumble outdoors and cross the courtyard to the observatory to check the clock. The weather performed as badly as expected and in mid-April he wrote to La Condamine to say that he had managed to grab only two observations and that he was having problems with the sector's eyepiece. La Condamine sent him a replacement, but the problems persisted.

Bouguer was unhappy with the sector's lack of 'solidity', and then his own frame started playing up. In a letter to La Condamine, he explained that he had been forced to suspend observations due to an attack of gout. La Condamine was mystified: Bouguer had not touched wine for four years. Then, when Bouguer resumed operations, the clock's main spring snapped, causing further delay. By June, Epsilon Orion had disappeared from the sky. After a brief rest in Cuenca, where he built a water clock, Bouguer trekked back over the five rivers to Tarqui, arriving at the observatory in time to have his teeth rattled by a series of earthquakes that rumbled and jolted intermittently for two weeks, disturbing the delicate settings of the zenith sector.

Up north, life was no less trying. By his own admission, La Condamine's role in the latest round of observations was 'not the brightest part of this joint work', but he was content to be 'useful' if it hastened the mission's departure for France. Determined that his own, relatively elementary observations would be conducted with the greatest accuracy, he built a new observatory. To reduce the effects on the telescope of heat differentials caused by mounting the instrument against an exterior wall, he erected a stand-alone three-foot-thick wall within the building. To this wall was attached a specially made copper frame for the fourteen-foot telescope and eyepiece, together with four screws for adjustment. In a refinement of earlier observatories, the window in the roof could be opened and closed remotely, from below, which avoided precarious excursions up and down the ladder. 'I was not only engineer of the machine,' he recalled later, 'but blacksmith, mason and roofer.' As an afterthought, he added: '& I realised that I had no vocation for the last trade.'

Like Bouguer, La Condamine suffered a run of misfortunes. One of the workmen he had employed to build the observatory spotted the small, leather-bound, silver-clasped notebook that contained observations and a handwritten table compiled in France by the esteemed topographer Abbé de la Grive. La Condamine had used the table for calculating angles during the triangulation survey, so the little book's job was partly done. The workman had thought that he was stealing a prayer book. The next misfortune was a violent storm that took off several roofs in the city, together with La Condamine's ingenious observatory window. Rainwater entered the telescope around the lens mounting and deformed the hairs of the micrometer. To repair the damage, the end of the telescope had to be dismantled by Hugo then re-soldered.

Progress was scarcely any better up at the Mira observatory, where Godin had fallen prey to three successive bouts of fever, the most recent lasting six weeks. Despite this, he had managed to collect a large number of observations through the months of May and June, before the weather broke in July. He stayed on at Mira till late August, but the news from Tarqui was not good. When the weather had been good at Mira, it had been bad at Tarqui, 'which meant,' as La Condamine wrote later, 'they had no corresponding observation'. The attempt to shoot Epsilon Orion simultaneously had cost another six months of pain and frustration.

For Godin, there was a momentary respite. Two weeks after he returned to Quito, Jorge Juan and Ulloa reappeared after a year away. Their summons to Lima had put them in the front line of the city's preparations for a British attack: converting two galleys and upgrading the city's defences. 'On our arrival at Quito,' remembered Ulloa, 'we made it our first business to join the French company, who were pleased to express a great deal of joy at our return.' The joy would not last.

As the surviving members of the mission mingled

and caught up with each other's news, it became clear to Jorge Juan and Ulloa that La Condamine had betrayed them and their homeland. The inscriptions on the carved tablets attached to the two monumental pyramids on the Yaruquí baseline had omitted Spain from the geodesic narrative. There was no mention on La Condamine's pyramids of King Philip V. Or of Don Jorge Juan or Don Antonio de Ulloa. Apparently, the Geodesic Mission to the Equator had been undertaken entirely by Frenchmen. Given what had happened in Quito's plaza, La Condamine was lucky not to have his knackers diced with a cutlass. The bonds between the mission's Spanish and French contributors had been forged on mountain-top and *páramo*. They had been close. Their lives had depended upon each other. But La Condamine had broken the bond. It was inconceivable that the monuments should exclude the Spanish names that had made the mission possible. A lawsuit naming La Condamine was filed and the mission was engulfed in a cloud of acrimony and court appearances. Only Bouguer and Morainville remained clear of the immediate fallout, locked to their zenith sector in faraway Tarqui.

It was a complicated disaster. The affront could be traced back to the wording that had been stipulated by the Académie des Inscriptions et Belles-Lettres, then carried to the Viceroyalty of Peru by Godin, the mission's leader, who had ceded responsibility for the pyramids to La Condamine. The French Academician had failed to understand how insulting the inscription would be to the Spanish lieutenants. For their part, Jorge

Juan and Ulloa were not masters of nuanced response. They were either on your side or coming at you with both barrels. The bust-up and lawsuit consumed three months that might have been spent up at the Mira observatory, concluding the astronomical observations.

By the time the row subsided, in December, it was too late. Jorge Juan and Ulloa were preparing to head up to the Mira observatory with Hugo when news arrived in Quito that the British had sacked and burned the port of Paita, only a couple of hundred miles south of Guayaquil. Commodore Anson was on the rampage. Nobody had expected a British fleet to round Cape Horn. The Pacific coast promised easy pickings. Guayaquil and Panama waited for the worst. For a second time, Jorge Juan and Ulloa were summoned to the front, this time by the *corregidor* of Guayaquil, who paid them to raise 300 armed men as they travelled from Quito to the coast.

As 1741 drew to a close, the mission was in a state of almost complete disarray. More than a year of astronomy had failed to fix the ends of the arc. The goodwill of Jorge Juan and Ulloa had been lost, and they had been called back to war. Godin had withdrawn from proceedings. Two of the mission were dead. The only sunny event at this time was a mission marriage. Jean-Baptiste Godin des Odonais, the young, redundant signal-fixer, had returned from his trading trip to Cartagena de Indias with money and with love in his heart. María Isabel de Jesus Gramesón had been born in Guayaquil but had moved inland at the age of five, when her father was awarded the post of *corregidor* of Otavalo, north of Quito. Isabel's Spanish mother claimed lineage back to the conquistadors, while her father was descended from solidly French stock. Isabel and Jean-Baptiste took their vows at the Dominican College in Quito and began dreaming of life in France.

Early in January 1742, Bouguer returned to Quito. He had been away for eleven months, a record incarceration in the observatory at Tarqui. Of them all, Bouguer was the best able to live with himself, apparently content to spend months in isolation, his rhythms dictated by the passage of the stars and his spare hours dedicated to his treatise on ship design. La Condamine once observed of his teetotal friend, that 'in his solitude', Bouguer pursued 'a very philosophical kind of life'.

Reunited, the two men reviewed their options. They may not have expressed it to each other, but this was the moment they had been waiting for. Godin was out of the picture. The mission had self-condensed to a team of two Academicians, backed up by an instrument-maker, a mapmaker and a draughtsman. They were a skilled, complementary quintet: the kind of team that might have been hand-picked back in Paris, had selection for the Geodesic Mission to the Equator been subjected to a rigorous process of test and interview.

They made a plan. Bouguer was convinced that the variations in their astronomical observations were attributable to faults in the twelve-foot Graham sector, an instrument that had been knocked about and repaired time and again. So the Graham sector would be put aside and La Condamine would become the second astrono-mer, using Godin's twenty-foot sector. And Hugo would be asked to go back to his workbench and build a new zenith sector for Bouguer.

In his humble *atelier*, Théodore Hugo assembled his tools and pieces of brass and copper and iron, the lenses and screws, the nuts and the bolts. And he began the painstaking process of building another new zenith sec-tor. This one would have an eight-foot radius.

He had only just started when – on 19 January – Ulloa unexpectedly appeared in Quito. He was filthy, ragged, exhausted and alone. And he had a story to tell. One month earlier, when he had left Quito with Jorge Juan, the two men endured an 'inconceivably fatiguing' nine-day journey by mule, foot and boat to Guayaquil, where they had participated in the council of war. It transpired that the British, having sacked Paita, had got wind of Spanish preparations to defend Guayaquil. Anson's fleet had sailed off north towards Panama. The council of war had agreed that either Ulloa or Jorge Juan could return to Quito in order to conclude their work with the French Academicians. The senior of the two, Jorge Juan, remained on defence duties in Guayaquil, relieving Ulloa to make the return trip inland. It was the worst time of year to travel. The rivers were in spate and the roads mired with mud. While crossing one of the rivers, two mules were swept away by the current, taking Ulloa's portmanteau with them. The muleteer survived by grab-bing the tail of one of the mules, but they were carried a quarter league downstream. Further into the mountains, the trail was in such an appalling state that Ulloa spent twelve hours covering half a league. It took him fifteen days to complete the journey. And as soon as he entered the city, he was handed orders from the Viceroy: 'I reached Quito,' he remembered, 'but had hardly alighted from the mules with the hopes of resting myself after these dangers and fatigues, when the president informed me . . . to hasten to Lima with all possible expedition.' Both he and Jorge Juan were needed again to help defend Peru from the British.

During his hectic two-day turnaround in Quito, Ulloa had a very unfortunate bust-up with La Condamine. Residual bad feeling about the inscription on the commemorative pyramids ignited during a collision between a pig-headed Frenchman and an exhausted Spaniard. Hearing that Godin was going to lend La Condamine the twenty-foot zenith sector, Ulloa asked his friend Valparda y la Ormaza – the crown attorney who had pulled the strings that led to the killing of the president's secretary – to impound the instrument, with instructions not to let anyone near it until he had returned from military duties. Without two sectors, Bouguer and La Condamine would not be able to conduct simultaneous observations from both ends of the arc. The mission was destroying itself from within.

Back in his *atelier*, Hugo was told that the old Graham sector would have to be repaired, after all. And of course he would still have to build a new eight-foot sector. It would take months.

Disintegration beckoned. Godin went off to look for lost treasure in the River Pisque. Bouguer was boiling about La Condamine's lawsuits and pyramid building. La Condamine resented Bouguer's recent habit of withholding observations. But as Hugo's work on the sectors neared completion in May, all three Academicians were invited by the Jesuits of the University of Tomás to present a thesis dedicated to the French Academy. It was rare for the trio to share a building, let alone a platform. And it was the last time they would be together in Peru.

That day, the Jesuits facilitated a partial *rapprochement*. La Condamine suggested to Bouguer that they return for one last time to Pichincha. Their first triangulation signal and hut had been placed on Rucu Pichincha, the 'Vesuvius of Quito'. Their friendship had been cemented during those interminable nights of ice and tempest on the spike-like summit. But neither of them had climbed its adjacent twin, Guagua Pichincha, three miles to the west. Guagua Pichincha was topped by an enormous, elevated crater that had been the source of several volcanic eruptions, most recently in 1660. Neither Academician had ever stepped inside a live volcano. To survey the interior of the crater, La Condamine would take his quadrant. They planned their adventure for mid-June, with the aim of being away for a week or so and then returning to Quito by the time Hugo had finished work on the new zenith sector. Like most La Condamine adventures, jeopardy was inevitable.

It went wrong from the start. On the morning of departure from Quito, the mules La Condamine had booked failed to appear. Bouguer became impatient and set off for the mountain with his own mules and the guide they had hired, leaving La Condamine on a Quito street with a mound of bedding and instruments, but no transport. With help from Quito's *alcaldes*, La Condamine left the city later that day, accompanied by two pack mules, a muleteer, two local men and a replacement guide. By sunset, he was stranded alone with his mule at the snow line, his companions having been unwilling to subject themselves to ravines, sub-zero temperatures and darkness. 'It was beautiful moonlight', remembered the Academician, 'and I knew the terrain . . . when I was suddenly enveloped by fog so thick that I absolutely lost myself.'

For hours, he blundered in the black fog, slipping and falling between thigh-high tussocks of sodden grass. Rain turned to sleet. Sometime after midnight, he curled his shuddering body beneath his cape with the mule's reins under his arm and waited for dawn. With the bitter cold draining all feeling from his limbs, he was forced at four in the morning to lever himself upright and piss on his feet to restore the circulation. At first light, 'bristling with frost', he descended to a farm, whose occupants lit a fire and brought him back to life. By evening, he was back in Quito, recruiting new guides. Later the following day – the 14th – La Condamine repeated the circuitous route south to Chillogallo, over the ridge to Lloa and then up the long valley to the declivity between the two volcanoes, where he found Bouguer waiting with the tent. It was a spectacular spot for camping. Framed by the two Pichinchas, they could see the silvery cone of Cotopaxi, a mere 30 miles to the south. Above their heads, Guaga Pichincha rose in a blinding, decapitated cone. So much snow had fallen that the bare slopes of pumice and lava had been smothered.

In the two days that Bouguer had been waiting for La Condamine to appear, he had explored the flanks of the volcano, looking for a route that would lead to a breach in the crater rim. But the eastern side of the volcano was riven with water-cut gullies and the deep snow impeded upward progress. On the 15th, the two men continued the search for a breach. On the 16th, they tried climbing a rib of exposed rock that appeared to lead straight up to the crater rim. Beyond the rib rose a steep snow slope. La Condamine pressed on alone, but became disoriented in sinking cloud. Bouguer's calls led him down to safety. On the 17th, they argued. Bouguer wanted to circle the volcano and try approaching it from the west; La Condamine wanted to return to the direct route: 'I offered myself to serve as a guide,' he said, as if this would reassure Bouguer that they were not climbing to their deaths.

Probing the deep snow with a pole and kicking steps as he climbed, La Condamine headed for the skyline. The men they had employed as guides turned back. The higher they floundered, the more certain became La Condamine that they were closing on the lip of the crater:

I approached cautiously a bare rock which overlooked all those on the crater. I turned it from the outside, where it ended with an inclined plane, with rather difficult access:

if I had slipped, I would have rolled on the snow 500 or 600 *toises* to the rocks, where I would have been very badly received. Monsieur Bouguer followed me closely, and warned me of the danger he shared with me: we were alone . . . Finally we reached the top of our rock, from where we saw the mouth of the volcano.

Breathlessly, the two Academicians peered into the crater, its sheer walls 'blackish and calcined' and at the bottom, 'the collapsed debris of the summit of the mountain during its conflagration: a confused heap of enormous rocks, broken and piled up irregularly on top of each other, presenting to my eyes a vivid image of the chaos of the Poets'. They could see no smoke or fumes rising from the crater, which was almost circular but for a breach in its western side. While the icy wind hacked at their faces and froze their feet and hands, La Condamine fumbled with his compass, taking bearings for a future map. Eventually, Bouguer succeeded in persuading his friend to descend to safety. They remained on Guagua Pichincha for another two days, attempting to find a viable route into the volcano. Before descending to the city, they were treated to the extraordinary spectacle of Cotopaxi suddenly erupting 'in a whirlwind of smoke'.

Back in Quito, they found that Hugo had finished work on the two sectors. A few days later, Bouguer sent his dismantled sector north to Cochasquí with his servant, who was instructed to prepare the observatory for the final round of simultaneous observations. Bouguer's impatience to conclude the astronomy was countered by

La Condamine's procrastination. He seemed to have an interminable list of tasks to complete before leaving the Viceroyalty. He was awaiting a decision on the 'pyramids affair' and letters were still being exchanged about the riot in Cuenca. The trial had produced a thousand pages of documentation, which La Condamine insisted on having copied and bound into a folio for transport back to France. Hugo was making him a new 'metal-rod' pendulum that he wanted to test before leaving Quito. La Condamine had also commissioned an elaborate commemorative bronze rule that he intended to leave in the city as a lasting monument. There were experiments to be concluded on the expansion of different metals of varying thicknesses. La Condamine wanted to expose each in turn to the sun, boiling water and snow. Incomplete projects and irresistible ideas bounced before his eyes day and night. It was as if he had opened a box of springs and lost control of their chaotic arcs. Urgently, he needed to raise money to pay for the return to Europe. Items no longer needed were put up for sale. The tent that La Condamine had bought in Saint-Domingue and used on countless mountains – and most recently on Guagua Pichincha – was erected in Quito's main square and sold for more than its purchase price to 'a gentleman who had a passion for hunting'. Precious time was devoted to planning the journey back to France. Characteristically, La Condamine had decided not to travel with Bouguer via Cartagena de Indias but to make his own way home by the most interesting (and hazardous) route available.

Bouguer became so impatient that he delivered an

ultimatum: the final observations would be cancelled unless La Condamine left Quito for the Tarqui observatory within two weeks. Unprepared to linger any longer, Bouguer departed for the Cochasquí observatory. Frantically, La Condamine concluded the adjustments to his zenith sector, then dismantled and packed the instrument into a specially built crate for transport to Tarqui. Carried by six porters and accompanied by Morainville, the crate left Quito on 4 August. The following day, La Condamine conscripted Verguin to help with a series of pendulum experiments, and on the 9th, he climbed Pichincha with his new metal-rod pendulum. He did not return to Quito for six days.

When La Condamine came down from Pichincha on 14 August he found that his room had been robbed. The heavy iron ruler on which he had marked the results of his experiments on the expansion of metals had gone. The ruler weighed 7 or 8 pounds and would have changed hands on a Quito backstreet for 7 or 8 ounces of silver.

The theft of the ruler cost more than 'the fruit of rather painful labour'. It was a valuable piece of metal. With the science in Quito concluded, the mission's armoury of instruments had become the currency of salvation. To pay for the long journey home, La Condamine and Bouguer needed every piastra they could muster. Three days after the theft, La Condamine cut a deal with a canon known to have 'a decided taste in machines'. For La Condamine, it was a painful negotiation, for the instrument he was flogging was his treasured three-foot Louville quadrant, the trusted, cumbersome beast he had lugged up the Esmeraldas river all those years ago. But it had served its purpose and the cost and trouble involved in transporting an instrument that needed two mules to move was too much. The canon handed over 1,500 livres for the quadrant, 600 more than La Condamine had paid in Paris. On behalf of Godin, La Condamine also sold the Graham clock, perhaps the most valuable instrument to have survived the Andes. It went to the rector of the University of the Dominicans in Quito. 'It is thus,' wrote La Condamine, 'that in a country where *les sciences & les arts* are not generally cultivated, a small number of people are the depositaries of this sacred fire.' In truth, they needed the cash.

Six days after descending Pichincha, La Condamine was congratulating himself that he had 'got rid of anything that might delay [his] march'. He had assembled the instruments, books and luggage that he would need at Tarqui and beyond. He had booked for the 20th a mule train that would carry his baggage to Cuenca. Allowing for the slow plod of the mules, La Condamine planned to leave Quito ten days later, in a *diligence*, one of the fast stagecoaches that operated along the Great Road to Riobamba. But La Condamine knew better than anyone that Peruvian plans had a habit of unravelling. On the morning of the 20th, he was reminded again that he was operating as a low-rent visitor. His booked mules were unavailable.

His last ten days in Quito were busy. There was an outstanding legal requirement to accompany a judicial officer to the pyramids at Yaruquí. La Condamine also wanted to have a final planning meeting with Bouguer, who was already out at the Cochasquí observatory. And there were goodbyes to be made in El Quinche. These agreeable duties could be linked on a short journey through the countryside north of Quito. It was a poignant excursion. With the official from the *audiencia*, he met Bouguer at the pyramids and the two Academicians were able to look for a last time along the sight lines of their first triangles, towards the signals on Pichincha, Tanlagua and Pambamarca, where La Condamine's timber cross was still visible as a tiny silhouette. Inevitably, La Condamine managed to concoct a minor adventure from thin air.

To save time, the two Academicians had left their bedding and luggage with mule drivers on the edge of a deep *quebrada* on the road to El Quinche, where they planned to spend a couple of nights with their friend, the priest José Antonio Maldonado. But by the time they left the pyramid at Caraburo, daylight was failing, and when they reached the *quebrada* they realized that it was too late to reach El Quinche before nightfall. La Condamine wanted to ride on to El Quinche, leaving the mules and luggage to catch up the next day. Bouguer declined the invitation to share a night-ride with his friend: he 'apparently remembered our adventure on Coto-paxi,' wrote La Condamine, and 'did not want to lose sight of his bed'. Characteristically, they failed to compromise, so La Condamine rode into the darkness and Bouguer unrolled his bedding on the edge of the *quebrada*, for a night under the stars.

The two days Bouguer and La Condamine spent with

'Docteur Don Joseph' at El Quinche allowed the Academicians to 'agree definitively' – as La Condamine put it – on the arrangements concerning the simultaneous observations at the two ends of the meridian arc. This would be their last chance to get it right. Bouguer was almost ready to start observing at Cochasquí. Allowing for travel time, La Condamine expected to be ready for observations at Tarqui within the next two or three weeks. This time round, the two men would be able to communicate their observations to each other through a hub at the Elén hacienda, halfway between Quito and Cuenca. They agreed to send data to each other every fortnight. On 27 August, they rode together for a league or so out of El Quinche and then their roads parted. Both knew that either they would meet again in France. Or never.

Back in Quito, La Condamine looked in horror at the splintered door to his study. He had been robbed again.

After bidding his adieus to Bouguer, he had ridden back to Quito that same day, the 27th. He wrote a report on the pyramids and delivered it to the office of the *audiencia*. He packed. And on the 31st, he reached at last 'that long-desired moment, and [I] was ready to ride, when the most cruel and unforeseen accident happened to me'.

The casket was gone. He had left it on the table. He had used it for the safekeeping of his most precious possessions. He had kept it with him in Quito rather than trust it to the mule train that had left already for Cuenca. Beneath its locked lid were his remaining treasures: ear-rings and nose-rings in copper and gold, drilled emeralds, some 'small, delicate works, of a very fine gold, found near the mouth of the river Sant-Iago'. In the casket, too, was the money he had hoarded for the journey home. And the all-important journals containing his observations and calculations.

It was the kind of catastrophe that makes the blood run cold. Of the myriad mishaps that had befallen him in Peru, none came close to the loss of his journals. They were all he had to show for the years away from the Academy. They were the passport he needed for a return to Paris. 'I admit,' he remembered later, 'that I was close to giving myself up to despair.'

Hollow with panic, he turned to Quito's *corregidor*, who published the same day an appeal for witnesses, with the promise that the owner of the stolen items would abandon the cash in return for the journals. That night, the heavens broke and the roofs of Quito drummed with rain. The water level in La Condamine's rain gauge rose above eight lines.

For forty long hours, La Condamine existed in a no-man's-land of numbness. At daybreak on 2 September he emerged from the door of his room to see a bundle resting beside the water fountain in the centre of the courtyard. Enveloped with relief, he opened the cloth to find his journals. Later, he realized that two small notebooks were missing. It took a while to work out why the thief had kept them. Inside one of the notebooks, La Condamine had written the heading 'Pitchincha' and, in the other, 'Coto-paxi'. Both mountains were rumoured to have undiscovered gold mines 'which many people imagined to have been the secret object of all our voyages in the mountains'.

The robbery, and yet another difficulty with Quito's judiciary concerning *l'affaire des pyramides*, kept La Condamine in Quito until 4 September. It was time at last to leave. He gave the rain gauge to Father Milanezio in the college of Jesuits.

As the walls of Quito shrank behind him, he acknowledged his mixed feelings:

> One can judge that this even [the theft], following all the unpleasant affairs which I had had in Quito for two years, was very likely to moderate the regrets I had in leaving a place singularly recommended for the gentleness and the equality of its climate, and which, after a stay of several years, I flatter myself that I have left some friends.

He would never see Quito again.

As he bumped along the Great Road for the last time, La Condamine had a number of calls to make. He had sent his quadrant ahead to an estate near Cotopaxi owned by his friend, the Marquis of Maenza, where he planned to pause in his journey and establish whether the recent eruption had melted sufficient snow to reduce the volcano's height. Not for the first time, Andean cloud rendered his scientific investigation 'useless' and the weary Academician rode on south after staying just one day.

Outside Ambato, La Condamine paused again, this time for a short detour to the hacienda of Pedro Vicente Maldonado. Of the Maldonado brothers, Pedro was closest to La Condamine. His work on opening the Esmeraldas road, his mapping projects and his loans of money to the mission won him gratitude and affection from the Academicians. And now La Condamine needed to finalize a plan that would place the two men in the same perilous boat. They agreed to meet at Lagunas, on the eastern side of the Andes, and then descend the River Amazon. Once at the coast, they would look for a ship bound for Europe. Whoever reached Lagunas first would wait for the other. It was a plan of casual flexibility.

From Ambato, La Condamine and Maldonado travelled on south together, to the Elén hacienda outside Riobamba, home of Maldonado's brother-in-law, José Dávalos, and of his multilingual daughters: the young women La Condamine referred to as *des Muses françoises*. There had been no deviation of devotion for the eldest of the three daughters. Four years after La Condamine had walked into the hacienda and become entranced by this family of francophones, María Estefanía was still set on becoming a Carmelite nun. The Elén hacienda was a place that had provided so much solace to the mission in the past. And now it was going to play a crucial role in the final act of this astronomical epic. Elén would be the communication hub – the *correspondance sûre* – during the forthcoming observations at Tarqui and Cochasquí.

Back on the Great Road, La Condamine was keen not to linger in Cuenca. Two weeks had passed since he had left Quito and his dallying was eating into the chances of snatching a simultaneous observation with Bouguer. And Cuenca was not a place with happy memories. Or of many friends. When he went to retrieve the trunk containing his clock and clothes, he found the lid open and half of the contents missing. It might have been a disastrous theft, but the pilferers had ignored the mysterious box of cogs and pendulums and instead helped themselves to La Condamine's clothing, prompting the owner to consider how fortunate he was that 'the thieves had needed shirts more than mathematical instruments'.

The loss of his shirts was a hint of what was to come. The miserable, isolated observatory at Tarqui had been prepared by Morainville, who had arrived a few days earlier. The twelve-foot zenith sector was in place, mounted on the wall, but La Condamine had trouble aligning the instrument on the meridian. Both the gnomon and iron clamps were missing. Bouguer had stripped them out and taken them for use at the Cochasquí observatory. Then La Condamine discovered that one of the feet of his quadrant had been stolen en route from Quito, together with its screws. He fashioned a replacement part out of wood. By October, he was still fiddling with the sector's pivot and trying to improve its rigidity. The weather was shit.

He had been at Tarqui for a month when alarming letters began to arrive from Bouguer, who had begun his observations at the end of August and now 'believed he had done enough, and that he renounced the simultaneous observations'. La Condamine had every reason to rant in his miserable billet. He wrote back, pointing out that during their first attempt at simultaneous observations, Bouguer had spent three months at Tarqui and yet failed to get a single result.

The impatience that M. Bouguer expressed to me added to mine. Never did a plowman, threatened by storms with losing his harvest, make more ardent wishes for a beautiful day than I made for a beautiful night: however, the rains ceased only to make room for fogs more annoying by their continuity than the rains themselves.

Bouguer agreed to stay on. Then the damp caused La Condamine's clock to start misbehaving. And earthquakes disturbed the sector. November seemed to be a continuum of rain, fog and tremors.

Then, at the end of November, the night airs cleared at Tarqui. And at Cochasquí. On the 29th, Bouguer observed the height of Epsilon Orion. And so did La Condamine. On the 30th, the night skies were clear at Cochasquí and at Tarqui. Unaware yet that they had achieved their first simultaneous observations, they kept at the sectors through December, observing whenever the skies allowed. At last, they had good news to share. The letters flowed through the Dávalos hub. The computations started. Bouguer got there first.

In late January 1743, he calculated that the length of one degree of latitude at the equator was 56,753 *toises*. Given that the lengths of one degree in Paris and at the Arctic Circle were 57,060 *toises* and 57,437 *toises*, they had proved beyond doubt that one degree of latitude was shorter at the equator than near the poles. The world was indeed Newtonian.

*

France and Spain could claim success for their joint scientific expedition. From the wings, Britain had provided the Newtonian script and meaningful props: the latest instruments. The shape of the Earth had been determined through international collaboration. Navigation and trade would be enhanced. But the mission had achieved much more. The complementary skills, competing interests and conflicting characters of the principal players had been catalysts of curiosity. They returned to Europe with new observations on subjects far beyond geodesy, from quinine, rubber and platinum to gravity, magnetism and aberration; from the speed of sound to the need for a universal unit of measurement. La Condamine published the first measured survey of a major Inca site. Jorge Juan and Ulloa investigated humanitarian crimes. The Geodesic Mission to the Equator was the model for a new form of expedition, motivated by the urge to *explore*.

Between the cordilleras, they had become known as *los caballeros del punto fijo*, 'the knights of the fixed point'. The quixotic measuring and observing were strange preoccupations in a world already known to its inhabitants. The visitors from Europe were collecting discoveries and stories. While Godin, Bouguer and Jorge Juan concentrated on the cosmos of numbers, La Condamine and Ulloa spun sentences from the world they could see, smell, hear, taste and touch. The diffusion of knowledge needed narrators as well as numerators. On their French passport, they were described as *Astronomes*, *Géomètres* and *Botanistes*. South America turned them into geographers. A decade spent exploring the coast and interior of the continent gave shape to an interconnected world in which erupting volcanoes and teeming rainforests shared space with gardens of farmland and chasms of thundering snowmelt. Villages occupied since the time of the Incas shared valleys with modern Spanish towns centred on plazas and churches. Balsa rafts, dug-outs and galleons rode the same sea lanes. The human geography of town and country was a nervous system of peoples whose capillaries reached across South America, the Caribbean, Africa and Europe: there were castes of inequality and injustice so extreme that Ulloa was moved for life. Five years before the mission sailed from Rochefort, the Academy had appointed its first geographer. In the years that followed the mission's return, geography emerged as a new science, absorbing from astronomy the disciplines of surveying and cartography and finding an audience eager to understand the transcendent power of the physical world. As he was packing to leave Lima for Cape Horn and home, Ulloa mused that his sovereign 'might not be totally disappointed in his generous views of promoting the useful sciences of geography and navigation'.

The Geodesic Mission to the Equator showed how a disparate bunch of human beings from different countries and backgrounds could use their collective brains to solve shared problems. They innovated. They combined ideas. They grasped that the dogged slog of incremental improvements would lead to a result. They proved future science.

14

Pierre Bouguer

After leaving Quito in February 1743, Bouguer became the first member of the mission to reach Europe. His arduous seven-month overland journey to Cartagena de Indias left him short of funds, but he boarded a ship for Saint-Domingue, where he sold his slave, released his servant and secured 2,000 livres for his onward passage to Europe. He crossed the Atlantic on an Irish-owned slave ship and landed in France in May 1744. The following month, he strode back into the Louvre, determined to capitalize on his nine-year absence. Bouguer's superior findings overshadowed the results brought back to France by the Arctic expedition. Not only had the Mission proved Newton right, but Bouguer was able to compute the actual curve of Earth's surface – an issue of great import to the navigators of the French Navy. Bouger became the Academy's celebrity scientist, writing widely on astronomy and mathematics, navigation and physics. In 1746, he published his long-delayed work on naval architecture, *Treatise of the Ship*. This was followed in 1749 by *The Figure of the Earth*, a detailed account of the Geodesic Mission to the Equator. Where Godin slipped into disrepute, Bouguer became the Academy's director. The reluctant sailor from Brittany has craters on Mars and the moon named after him. Meteorologists remember him through the term 'Bouguer's halo', the phenomenon he observed on Pambamarca when his backlit, rainbow-crowned image was projected onto cloud. In geology, a 'Bouguer anomaly' relates to variations in the Earth's gravitational field caused by differences in the density of underlying rocks. In Brittany, a bronze statue of Le Croisic's most famous son looks across the harbour waters with a quadrant at his hip and a triangulation map in his hand. Teetotal in Peru and unmarried in Paris, Pierre Bouguer was wedded to science. Early in 1758, he was felled by amoebic dysentery.

Charles-Marie de La Condamine

After successfully completing his astronomical observations with Morainville at the southern end of the chain of triangles, La Condamine followed his plan to take the hard route home. With one of the slaves he had inherited from Seniergues, he left Tarqui on 1 May 1743 and took the road to Loja, where he collected cinchona saplings and evaded assassins seeking revenge for the imprisonment of Cuenca's *alcalde*. Mountain trails, fords and swaying liana bridges led him through the cordilleras towards the headwaters of the Amazon. On a tributary of the Marañón, La Condamine commissioned the construction of a balsa raft, and it was on this precarious vessel that he was dashed repeatedly against rocks while shooting the notorious eddies and rapids of Pongo de Manseriche. He timed his descent of the rapids at fifty-seven minutes and emerged 'upon a fresh water sea, surrounded by a maze of lakes, rivers and canals, penetrating in every direction the gloom of an immense forest'. At Lagunas, La Condamine achieved the planned rendezvous with Maldonado and they continued down the River Amazon in a pair of forty-foot dug-out canoes, recording physical and human geography, making maps, collecting samples and conducting experiments. On 27 September 1743, they reached the coastal port of Belém. War delayed a return to France, and La Condamine finally reached Paris in February 1745. Back in the Louvre, he found the Academy enthralled by Bouguer. While Bouguer stuck to the hard science, La Condamine became the popular mouthpiece for the mission's extraordinary adventures in South America. Voltaire wrote of his friend dropping by for *café au lait* on the way to the Academy and La Condamine wrote extensively about his experiences and findings in Peru. *Journal of the Voyage made to the Equator* and *Measure of the First Three Degrees of the Meridian in the Southern Hemisphere* were published in 1751. Ahead of his time, he also wrote a paper proposing that all nations should adopt a standard unit of length, but it would be another half-century before the metre was formally defined in French law as one ten-millionth of the distance between the North Pole and the equator. La Condamine married in 1756. In 1774, he insisted on being the trial patient for a new form of hernia surgery and died of blood poisoning. He bequeathed his papers to his old friend Maupertuis.

When Jorge Juan and Ulloa returned from their military duties in January 1744, they found Godin alone in Quito. Between January and May 1744, the three men extended the triangles northwards to their Mira observatory, then used the impounded twenty-foot zenith sector to complete the astronomical observations that would fix the northern end of the chain. Unlike Bouguer and La Condamine, the trio did not attempt simultaneous observations from both ends of the chain of triangles. Their final computed figure for the length of one degree of latitude was 56,767 *toises*, which was 14 *toises* (90 feet) longer than the figure determined by Bouguer and La Condamine. It was a remarkably small discrepancy and one that substantiated the conclusion that the world was indeed oblate. Before their return to Europe, the two lieutenants duplicated their findings then boarded separate ships in case one of them should fail to survive the voyage. The two vessels sailed from Callao in October 1744, bound for Cape Horn and the Atlantic. Jorge Juan reached Madrid early in 1746. Promoted to captain, he was instructed by the Spanish Secretary of State, the Marquis of Ensenada, to submit with Ulloa an account of the expedition at government expense. Jorge Juan's contribution was a volume covering the scientific aspects of the ten-year expedition. In 1749, Jorge Juan was sent as a spy to England. Travelling as 'Mr. Josues', he collected information on naval construction and armaments and communicated his findings in numerical code to Ensenada. His subsequent role as a royal troubleshooter took him into defence and engineering, mining and irrigation. In 1767, he was appointed ambassador to Morocco and spent his last years back in Madrid as head of the Royal Seminary of Nobles. He remained unmarried and died in 1773. With time, his life achievements led to repeated recognition: his name has been painted to the stern of two successive twentieth-century Spanish naval destroyers; there is a Calle Jorge Juan in his hometown of Valencia and his portrait has graced the reverse side of a 10,000-peseta banknote.

Antonio de Ulloa y de la Torre-Guiral

As the younger of the two Spanish lieutenants on the mission, Ulloa had been junior to his friend in most matters, and when the two men left Callao in 1744 Ulloa was on board the smaller of the two merchant ships, the slow and leaky *Notre Dame de la Délivrance*. After safely rounding Cape Horn, *Délivrance* sought shelter from the British in Louisbourg, unaware that the port had been captured. To avoid sensitive information falling into enemy hands, Ulloa threw many of his papers overboard. He crossed the Atlantic as a prisoner of the British, who jailed him near Portsmouth. Eventually, the president of the Royal Society in London heard that a member of the Geodesic Mission to the Equator was languishing in a Hampshire jail. Ulloa was released and reached Madrid several months after Jorge Juan, in July

1746. Two weeks after being reunited, the two men submitted their book proposal. Four of the five volumes of *A Voyage to South America* were written by Ulloa. The geographical page-turner was published in 1748 and translated into several languages. As soon as Ulloa had finished this masterwork, he was ordered by the government to submit 'a confidential report on the civil and political government of these kingdoms' in South America. In the report, Ulloa launched an unrestrained assault on the ills of colonial Spain: the iniquity of the *mita* system, the tyranny of the *corregidores*, the exploitation of villagers by priests and the social abrasion between the European Spanish and the established settlers. Had Ulloa's secret report been published in the eighteenth century, it would have sent shockwaves through Spain. *Secret News about America* was eventually published in 1826 and remains one of the landmark works to emerge from the Geodesic Mission to the Equator. Ulloa's devotion to writing, reading, research and experimentation was interrupted by a posting back to Peru to eliminate corruption in the mercury mines and a posting to New Orleans as governor of Louisiana. Twice, he was ordered back to sea as commander of an Atlantic squadron. In Cádiz, his curiosity led him into electricity and artificial magnetism, solar reflection and blood circulation in fish. He also worked on permanent inks, bookbinding and metal letterpress, and he introduced finer wools to Spanish weavers. An English clergyman who visited Ulloa in the late 1780s left a touching description of the scene: 'This great man, diminutive in stature, remarkably thin and bowed down with age, clad like a peasant' occupied a room that measured 20 by 14 feet, in which were 'dispersed confusedly, chairs, tables, trunks, boxes, books, and papers, a bed, a press, umbrellas, clothes, carpenter's tools, mathematical instruments, a barometer, a clock, guns, pictures, looking-glasses, fossils, minerals and shells, his kettle, basons, jugs, American antiquities, money . . .' Ulloa kept writing to the end of his life. *Conversations with His Three Sons in the Naval Service* was published in 1795, the year he died.

Louis Godin

By the time the mission disbanded in 1744, its one-time leader had lost the will and the means to compute his own definitive figure for the shape of the Earth. He was in debt and his companions were gone. In Paris, Maupertuis accused him of being 'dishonoured by all the embezzlement' and of sowing 'implacable hate and discord' among his companions, and finally of being so 'embarrassed and fearful' that he had taken refuge in Peru. In December 1745, Godin suffered the ignominy of being expelled from the French Academy of Sciences. By then, he was in Lima, where he had taken the well-paid chair of mathematics at the University of San Marcos, a post made available by the death of the mission's long-time friend and mentor Pedro de Peralta y Barnuevo. Two years later, the city was devastated by an earthquake that killed thousands and ended Godin's short university career. Subsequently, he played a significant role as a surveyor and town-planner in rebuilding the city. Godin kept to himself the mass of observations and journals he had accumulated during his years with the mission in Peru and never submitted a formal record of the expedition. In 1751, he returned to Paris, where he was reunited with his wife, Rose Angélique. Encouraged by Jorge Juan and Ulloa, the couple moved to Spain, where Godin was appointed director of the Naval Academy in Cádiz. The Academy of Sciences eventually readmitted Godin but, four years later, he died of a stroke, aged fifty-eight.

Joseph de Jussieu

By the time the mission finally fragmented, Jussieu was peso-less and ill. It took the melancholy doctor two years to accumulate sufficient funds to pay for the voyage home, but he was kept in Quito by a smallpox epidemic. In 1747, he travelled to Lima, where he was reunited with Louis Godin. The two of them undertook a journey to Lake Titicaca, where Jussieu studied aquatic birds. After leaving Godin, Jussieu was drawn to the silver-mining town of Potosí, where he became involved in the fate of miners poisoned by mercury. He stayed four years, returning to Lima in 1755 depressed and exhausted. He never fully recovered. In 1771, after thirty-six years away and following repeated pleadings from his family, he travelled home to France, where he spent his last eight years despondent and housebound, being nursed by his brother Bernard and nephew Antoine-Laurent. He never published his extensive botanical findings and most of his collections were lost during his travels.

Jean-Joseph Verguin

In the expedition's published accounts, the roles played by the four French specialists – Jussieu, Verguin, Morainville and Hugo – were under-recorded. Verguin, the mapmaker–engineer, was a versatile, practical contributor to the mission's evolving projects, the kind of individual who becomes indispensable as challenges multiply. Of his many contributions, one of the most significant was the manuscript map he completed in Quito in July 1738, which provided the surveying teams with a record of what they had achieved in the first season of triangulation. A subsequent map drawn by Verguin showed the full chain of triangles stretched like a lattice between baselines 200 miles apart. After the mission's work was completed, Verguin was delayed in Quito by illness and did not leave Peru until 1745, two years after Bouguer. When he eventually reached France and his home in Toulon, he found that his wife had died and his two children were being looked after by their grandmother. Verguin remarried and returned to his role as engineer for the port of Toulon. His meticulous 1752 plan of Toulon's arsenal and docks is an image of order that would not have been apparent at the

Caribbean and South American harbours he visited during his decade away. For his contribution to the Geodesic Mission to the Equator, he was nominated as a corresponding member of the Academy of Sciences. He died aged seventy-five in April 1777.

Jean-Louis de Morainville

Taken along as draughtsman and artist, Morainville's contribution to the mission went far beyond his initial brief. He was a key member of the triangulation teams and worked with La Condamine on astronomical observations. He helped La Condamine build the fateful Yaruquí pyramids and illustrated the monuments for publication. Perhaps his crowning monument was the *Plan de Quito*. Morainville compiled the map in 1741; it was the first detailed plan of Quito published and it remains a unique snapshot of a mid-eighteenth-century Spanish colonial city. Morainville is also remembered for making the earliest detailed drawing of the cinchona tree. The draughtsman and artist who had been such an invaluable member of the team from its inception through to its dispersal was abandoned by the Academy. Attempts to claim compensation from Maurepas failed. Morainville never saw France again, or his wife. He found work as an artist and architect and, in around 1765, aged about fifty-eight, he was helping to repair the church in Sicalpa, on the edge of Riobamba, when he fell to his death from scaffolding.

Théodore Hugo

This fascinating, scarcely documented member of the mission has no known birth date but would probably have been in his late twenties or early thirties when the mission sailed from Rochefort. His trade as a clock-maker would have required him to be an expert micro-mechanic, with a wide range of metalworking skills, from turning on a lathe to thread-cutting and casting. To La Condamine, he was 'proud Hugo our watchmaker' and '*Sieur* Hugo, our instrument engineer'. The *horloger* whose technical expertise made the mission's work possible was largely written out of the published accounts and eventually abandoned in Peru by the Academy. Hugo, too, unsuccessfully tried to claim funds from Maurepas. He eventually turned his back on his trade and opened a tile-making business in Quito and married a local woman with whom he produced many children. He died in around 1781.

Jean-Baptiste Godin des Odonais

Following his loyal and indefatigable service as a 'signal carrier' and astronomical assistant for the mission, Godin des Odonais took up the textile trade and settled with Isabel Gramesón in Quito, where their first child was born. They had plans to save money for the voyage to France, but fate kept tripping them up. Their baby died and neither the textile trade nor an attempt to make money from tax collecting led to enduring success. When an epidemic swept Quito in 1744, they moved with Isabel's extended family to the healthier airs of Riobamba. A second and then a third child died and then, in 1748, Godin des Odonais received a letter from France that had been written eight years earlier. In it, he learned of the passing of his father and of the wish of his family that he return home to St-Amand-Montrond. He devised a plan as daring as that of his mentor and hero La Condamine: he would descend the Amazon to the Atlantic, and, having checked the route's viability, return up the Amazon to collect Isabel, who was again pregnant. Then they would descend the Amazon together and sail to France. In March 1749, Godin des Odonais set off down the Amazon, following the route taken by La Condamine six years earlier. After seven months, he reached the Atlantic, but was unable to return upriver due to lack of funds and obstruction from the Portuguese authorities. Repeated attempts to raise money and to secure permissions failed. Nineteen years after they parted, Isabel set off down the Amazon with a group of forty companions in an attempt to reach the coast, where Godin des Odonais was still waiting. Most of her group died on the journey and she was forced to trek alone through the Amazon rainforest until rescued by villagers, who nursed her back to health. She was reunited with her husband on 18 July 1770. By June the following year, they were in France. The couple retired to Jean's family home in St-Amand-Montrond. Jean never managed to get his book of Quechua grammar published, but La Condamine included in a later edition of his own book, a 7,000-word account of Isabel's amazing journey, written by Jean. Through the efforts of La Condamine, Godin des Odonais was eventually awarded a pension for his services as 'official geographer to the king'. He died in 1792, aged seventy-nine, and Isabel died seven months later, aged sixty-five. In 2004, their story became a bestselling book by Robert Whitaker: *The Mapmaker's Wife, A True Tale of Love, Murder and Survival in the Amazon*. Today, St-Amand-Montrond is twinned with Riobamba.

Jean Seniergues

Murdered at thirty-five, the surgeon exerted a greater leverage on the mission in death than in life. His money-making excursion to Cartagena de Indias made him an early absentee from the expedition, and his escalating lust for riches diminished his contribution. Following the bloody 'Cuenca affair', La Condamine's obsessive pursuit of the murderers intermittently distracted the Academician for the last three years of geodesy. Larrie D. Ferreiro, whose outstanding book *Measure of the Earth* is the most comprehensive account of the mission, wondered whether Seniergues was suffering from early-stage paranoid schizophrenia. The belligerence and avarice that the surgeon was susceptible to cannot have been helpful to his more sensitive friend, Joseph de Jussieu.

Jacques Couplet-Viguier

Young Jacques was one of the expedition's random victims. According to La Condamine, he was the 'most robust' of them all. His presence on the first recce out of Quito in search of a baseline was a measure of his eagerness to contribute to the mission and to see and feel an exciting new land. He was seventeen when they sailed from France and eighteen when he died at Cayambe of 'putrid fever'.

Grangier

'I do not enumerate how many of our servants, white as well as of colour, who died in the course of our travels; two of them of a violent death': Charles-Marie de La Condamine, 1751.

Two hundred years after these words were written, they seem incomprehensible. The twelve men from France and Spain whose expeditionary activities were best recorded were accompanied by a changing cast of over a dozen servants and slaves. Of these *domestiques*, the only one who can be detected after the mission dispersed is the servant bought by Bouguer in Saint-Domingue in 1735.

Grangier endured the entire expedition and accumulated skills that included surveying, mapmaking and astronomical observation. He worked with Bouguer in the 'secret' observatory on the outskirts of Quito and at the Cochasquí observatory, too. When the observations were completed in 1743, he accompanied Bouguer on the long overland trek to Cartagena de Indias and onward by ship to Saint-Domingue, where he was released and employed as a royal surveyor for the French colony. There remains the intriguing possibility that Grangier was still alive when the plantation slave François-Dominique Toussaint-Louverture led the revolution that overthrew the French colonial rule of Saint-Domingue and established the world's first black republic.

Dear Reader,

Thank you. Where would authors be without you?

You'll know by now that this is a story rather than an academic work of reference. I wanted to swap my desk for a deck and sail with scientists in search of solutions. Written under siege from the pandemic, it's been a lockdown journey of discovery that has taken me to the highs and lows of the New World: the loft and the kitchen. For a while, I was able to ride my bicycle to and from the London Library, returning each time loaded like a galleon with bibliographic booty.

Most of my raw material has been quarried from books, journals and online libraries. The loft released a box-file containing the notebook, maps and assorted receipts and papers I'd cached after travelling by foot, bus, dug-out and steam train through Ecuador in 1989. Many of the places in this story sprang back to life when I revisited their scribbled descriptions in that forgotten notebook. I was travelling with a paperback copy of *The Conquest of the Incas*, by John Hemming, making sense of the ebb and flow of conquistador and Inca through that spectacular land. I was also travelling with a beautiful young woman and it was with a view to proposing marriage that I suggested a romantic mountain hike along an old Inca trail in the mountains above Alausi. Like the Academicians of the 1730s, we were unprepared for the severity of Andean mountaineering and spent our first night at over 4,000 metres caked in ice inside a pair of bin bags bought at a market stall. The following day, we traversed the long, exposed crest of Cuchilla Tres Cruces, less than a couple of kilometres from the signal where Bouguer, La Condamine and Ulloa had huddled more than two centuries earlier, their tent shredded and its poles snapped by ferocious storms. My photographs show blue horizons pierced by silvery volcanoes and rugged ridges that had been the settings of this story. After three days of wandering in the Andes, slightly lost, we came down to the ruins of Ingapirca, where La Condamine had conducted his ground-breaking survey of an Inca site.

So researching this book has been an exercise in narrative archaeology, excavating and examining fragments of information from a multitude of sources and then arranging them into an order that remains true to its subjects. I had wondered whether to include in the finished book the footnotes that accumulated as I scaffolded the research. Most are matter-of-fact notes detailing the source of a quote or explaining how I reached the conclusion that the mission's 'Sinasaguan' is likely to be the modern peak of Naupan. Some of the footnotes record authorial detours into arcana such as the efficacy of pepper and gunpowder-infused lemons as a cure for gangrene of the rectum. Contented hours were devoted to defining the word '*canonnière*', which – according to the 1883 edition of *Boy's Own Annual* – was a small military tent. I am a fan of footnotes, but at the latest count this book has 796 of them, amounting to nearly 28,000 words and I'm keen that this story should not be cluttered with distracting personal enthusiasms, so I've left them out. Most of the quotes I've used come from the works of the four mission members who wrote books and papers about the expedition.

La Condamine shines as the anecdotalist. Of his two principal books, *Journal of the Voyage Made to the Equator* (1751) is a week-by-week mosquito-on-the-wall account of this frequently chaotic eighteenth-century expedition. The partner volume, *Measure of the First Three Degrees of the Meridian in the Southern Hemisphere* (1751), is his description of the triangulation, astronomy, instruments and calculations deployed during the mission's decade in South America. Pierre Bouguer's more rigorous *The Figure of the Earth* (1749) opens easily enough with a travelogue that includes a section on his coastal antics with La Condamine and an examination of volcanoes and earthquakes. But the core of the book is a mathematical exposition of the geodesic investigations, with ample consideration given to the thrills of trigonometry, astronomy and the finer points of quadrants, pendulums and zenith sectors. *Historical Relation of the Voyage to South America* (1748) by Jorge Juan and Ulloa wins the expedition award for the most thorough geographical descriptions of these diverse lands and peoples. It is from Ulloa's pen that we learn of Quito's iniquitous caste system and the amphibious Pacific 'bird-child', better known today as the penguin. Jorge

Juan and Ulloa's second volume, *Astronomical and Physical Observations*, describes the geodesy. All five of these primary sources can be found online. Abridged sections of books by La Condamine, Bouguer and Ulloa, translated into English, can be found in the 14th volume of *A General Collection of the Best and Most Interesting Voyages and Travels in All Parts of the World* (1813), edited by John Pinkerton. The eighteenth-century supremacist undercurrent that surfaces now and again in the mission memoirs is best countered by reading the secret dossier compiled by Jorge Juan and Ulloa. An exposé of corruption and cruelty in Spanish South America, *Secret News* or *Information about America* was not published until 1826 and exists today in an English edition (*Discourse and Political Reflections on the Kingdoms of Peru*, 1978), edited and translated by John J. TePaske). Of the other members of the mission, Jean-Joseph Verguin's contributions have endured in the maps that the engineer-mapmaker left for posterity. A collection of them can be viewed on the Bibliothèque Nationale de France (BNF) website. The BNF are also custodians of an extraordinary digital vault containing the annual editions of the *Histoire de l'Académie Royale des Sciences*. Each edition is divided into a shorter, opening section of *histoires*, summarizing the latest scientific research, followed by the more comprehensive *mémoires* for the year. A dip into the 1735 edition reveals the pendulum reports sent by Godin and La Condamine from Saint-Domingue, and an exploded diagram of the instrument. The annual editions of the Academy's *La Connaissance des temps* (*Knowledge of the Times*) are also on the BNF website. For international banter of the 1730s and 40s, Volume 1 of Rigaud's *Correspondence of Scientific Men of the Seventeenth Century* (1841) is an absorbing browse and includes letters from Newton and Huygens, as well as the epistolary eruption from Bouguer and La Condamine in Quito, griping about Godin to Edmond Halley in London. La Condamine's jibe about the fat-jowled Parisian poet Sinetti can be found in *The Complete Works of Voltaire 87, Correspondence and Related Documents, III, May 1734 – June 1736*, edited by Theodore Besterman (1969).

The Geodesic Mission to the Equator has been revisited by many modern writers. Larrie D. Ferreiro's magnificent book, *Measure of the Earth: The Enlightenment Expedition That Reshaped Our World* (2011) marshals new research and covers mission sub-plots such as the War of Jenkins' Ear. For a romantic romp in the footsteps of mission-member Jean-Baptiste Godin des Odonais, read *The Mapmaker's Wife: A True Tale of Love, Murder and Survival in the Amazon* (2004), by Robert Whitaker. Neil Safier's more academic take, *Measuring the New World: Enlightenment Science and South America* (2008) is excellent on the currents of knowledge flowing to and fro across the Atlantic. Two important secondary sources focus on the mission's contributions to geodesy: *The Quest for the True Figure of the Earth: Ideas and Expeditions in Four Centuries of Geodesy* (2005), by Michael Rand Hoare, includes chapters on early geodesy and on the Arctic expedition led by Maupertuis. James R. Smith's *From*

Plane to Spheroid: Determining the Figure of the Earth from 3000 B.C. to the 18th Century Lapland and Peruvian Survey Expeditions, was published in 1986 to mark the 250th anniversary of the expeditions to Peru and Lapland. Mary Terrall's book, *The Man Who Flattened the Earth: Maupertuis and the Sciences in the Enlightenment* (2002), follows the life of the Academician who threatened to be the mission's nemesis. Joseph Townsend's touching description of an elderly Ulloa surrounded by a lifetime of treasures in Cádiz, appears in Arthur Whitaker's 'Antonio de Ulloa', published in *The Hispanic American Historical Review* of May 1935. In 'Charles-Marie de la Condamine's Report on Ingapirca and the Development of Scientific Field Work in the Andes, 1735–1744' (*Andean Past 2,* 1989), Monica Barnes and David Fleming recast the Academician as 'the earliest observer to have surveyed and analyzed a pre-Hispanic site in the Americas from the viewpoint of someone interested in historical interpretation.' Bouguer receives a generous rejuvenation in *Earth Sciences History, Vol 29,* (2010), where John Smallwood re-examines the hard-won pendulum observations in 'Bouguer Redeemed: The Successful 1737–1740 Gravity Experiments on Pichincha and Chimborazo'.

There are a maze of glittering seams ready to release contextual nuggets. Among the labyrinths of distraction I've enjoyed are Nicholas Cronk's *Voltaire, A Very Short Introduction*, 2017, which provides a train journey's worth of insights into one of the mission's more influential friends and spectators, while Ian Davidson's *Voltaire, A*

Life (2010) is a thoroughly enjoyable full-length biography. An illuminating essay by John C. Rule on Minister for the Navy, Jean-Frédéic Philippe Phélypeaux, Count of Maurepas, appears in the *Louisiana History* edition of Autumn 1965. The story of Taita 'Father' Buerán and the Cañari came from Rosaleen Howard's chapter ('Why do they steal our phonemes? Inventing the survival of the Cañari language . . .)', in Carlin, E., van de Kerke, S., (eds), *Linguistics and Archaeology in the Americas: The Historization of Language and Society* (2010). Nobody has portrayed the sad saga of the Incas more vividly than John Hemming in *The Conquest of the Incas* (1970). For transatlantic geopolitics and for the dodgy dealings of the South Sea Company, I turned to Adrian Finucane's *The Temptations of Trade: Britain, Spain, and the Struggle for Empire* (2016). Geoffrey Walker's *Spanish Politics and Imperial Trade, 1700–1789*, published in 1979, covers the traffic (both regulated and illicit) passing through Cartagena de Indias, Portobelo and Panama. Tamar Herzog's *Upholding Justice: Society, State, and the Penal System in Quito (1650–1750),* uncovers the brutal presidential tenure of Alsedo in Quito. Mark Honigsbaum's *The Fever Trail–The Hunt for The Cure for Malaria* (2001) leads the way into the cinchona forests of Loja. For the Cassini family and the reasons that French mapmakers of the eighteenth century regarded themselves as world leaders, there is no better source than Chapter 9 of Jerry Brotton's *A History of the World in Twelve Maps* (2012). In Michael Heffernan's paper 'Geography and the Paris Academy of Sciences: politics and patronage in early 18th-century

France', published in volume 39 of *Transactions of the Institute of British Geographers* (2014) we are guided through the Louvre archives and shown how the expeditions to Peru and Lapland contributed to the 'new science of geography establishing, arguably for the first time, the conceptual terrain on which the modern discipline would later be enacted'.

There are hundreds of books and papers on scientific instruments and geodesy. A reliable author to start with is the inestimable Eva Germaine Rimington (E.G.R.) Taylor, whose essential works on navigation include *The Geometrical Seaman: A Book of Early Nautical Instruments* (1962) and *The Haven-Finding Art, A History of Navigation from Odysseus to Captain Cook*, (1956). From the incomparable stacks of the London Library, I borrowed a copy of Colonel A. R. Clarke's *Geodesy* (1880), which includes a section on Bouguer's attempts during the storms on Pichincha to measure with a pendulum the effects of gravity at high altitude.

For authentic Andean mountaineering, I turned to Edward Whymper's *Travels Amongst the Great Andes of the Equator* (1892). Although it was published over a century after the departure of the mission from the Viceroyalty of Peru, Whymper's forays to – and above – the snow line were little different to those of La Condamine and his fellow mountaineers. For those of you who enjoy maps and mountaineering, a number of sources can be combined to retrace the mission's escapades. Try matching Verguin's sketch map of 1746 (it's on the BNF website) with the online 1:50,000 topo maps prepared by the Instituto Geográfico Militar (IGM), Quito, Ecuador. Then compare these to satellite images on Google Earth. One of the most startling topographic transformations has been the growth since 1746 of Quito, from a compact, 1-mile grid to a 30-mile sprawl. The long, level plateau selected for the baseline at Yaruquí has become the main runway of Quito's new international airport.

Which reminds me to mention the vexing issue of place-spellings. The individual members of the expedition deployed a confusing variety of place-names, especially with regard to mountains, most of which had not been named on available maps. The rule I have followed for this book is to use place-name spellings from the *Times Comprehensive Atlas of the World*. Where a place, mountain or river is too small to appear in the *Atlas*, I have turned to the spellings used on the 1:50,000 IGM maps. One place-name can lead to especial confusion: the 'Riobamba' of 1735 has not just changed name, but location. Following a devastating earthquake in 1797, the town was moved 20 kilometres eastward. The 1730s 'Riobamba' that the mission visits in this book, is now known as Cajabamba. For distance measurements, I've use *toises*, feet and miles, which are truer to the units that were in use at the time. Metres and kilometres had yet to appear as universal units, although their eventual adoption was due in part to La Condamine.

Writing this book has been a lot of fun. I'm a geographer at heart, and this is a geographical adventure story. It has been a lockdown lifesaver. But it would not have happened if an e-mail had not popped onto my screen just before Covid struck.

The idea for this book came from Dan Bunyard, Publishing Director and Head of Non-Fiction at Michael Joseph. It was an irresistible story: a cosmic quest with sailing ships, maps, mountaineering, extreme triangulation, mutinies, murder and a cinematic cast. On the afternoon of my first meeting with Dan, lockdown hit London, but already, I was committed. Thank you, Dan, for such an entertaining commission. And thank you, Jim Gill at United Agents, for your continuing wisdom, encouragement and for a timely walk through the invigorating borderlands of Alfred's England. My geography undergraduate friend Martin Goodchild read an early draft and listed in excruciating detail the many inconsistencies and inappropriate asides. All the surviving errors are of course mine.

As this book crept closer to the light of day, I've been able to share its journey with the wonderfully cheerful, efficient team at Michael Joseph: Senior Editorial Manager Bea McIntyre, Editorial Assistant Aggie Russell, map-commissioner Fran Monteiro and copy-editor Sarah Day. With the expertise of Gaby Young and Sriya Varadharajan in the Publicity department and Sophie Shaw in Marketing, *Latitude* is in good hands. Thank you all.

For permissions to use the images in this book please see Picture Credits on page 246. Except for the view of Martinique, all of the black-and-white illustrations come from the works of Bouguer, La Condamine, Jorge Juan and Ulloa.

In closing, I'd like to dedicate this book to everybody who cares about science.

Nick Crane, London, 2021

Picture Credits

Inset 1

Image 1: Bridgeman Images
Image 2: Alamy Stock Photo
Image 3: Bridgeman Images
Image 4: Maurice Quentin de la Tour (French, 1704-1788). Portrait of Charles-Marie de la Condamine, 1753. Pastel on paper, 19 1/4 x 17 1/4 in. Frick Art & Historical Center, Pittsburgh, 1970.40.
Image 5: Alamy Stock Photo
Image 6: Alamy Stock Photo
Image 7: Alamy Stock Photo
Image 8: Mary Evans Picture Library
Image 9: Gallica – The Bibliotheque Nationale de France Digital Library
Image 10: Gallica – The Bibliotheque Nationale de France Digital Library
Image 11: Bridgeman Images
Image 12: Gallica – The Bibliotheque Nationale de France Digital Library
Image 13: Alamy Stock Photo
Image 14: Bridgeman Images
Image 15: Gallica – The Bibliotheque Nationale de France Digital Library
Image 16: Alamy Stock Photo

Inset 2

Image 17: Gallica – The Bibliotheque Nationale de France Digital Library
Image 18: Gallica – The Bibliotheque Nationale de France Digital Library
Image 19: Gallica – The Bibliotheque Nationale de France Digital Library
Image 20: Gallica – The Bibliotheque Nationale de France Digital Library
Image 22: Bridgeman Images
Image 23: Gallica – The Bibliotheque Nationale de France Digital Library
Image 24: Mary Evans Picture Library
Image 25: Bridgeman Images
Image 27: Alamy Stock Photo
Image 28: Alamy Stock Photo
Image 29: Bridgeman Images
Image 30: Gallica – The Bibliotheque Nationale de France Digital Library
Image 31: Bridgeman Images

Index

He just wanted a decent book to read ...

Not too much to ask, is it? It was in 1935 when Allen Lane, Managing Director of Bodley Head Publishers, stood on a platform at Exeter railway station looking for something good to read on his journey back to London. His choice was limited to popular magazines and poor-quality paperbacks – the same choice faced every day by the vast majority of readers, few of whom could afford hardbacks. Lane's disappointment and subsequent anger at the range of books generally available led him to found a company – and change the world.

'We believed in the existence in this country of a vast reading public for intelligent books at a low price, and staked everything on it'
Sir Allen Lane, 1902–1970, founder of Penguin Books

The quality paperback had arrived – and not just in bookshops. Lane was adamant that his Penguins should appear in chain stores and tobacconists, and should cost no more than a packet of cigarettes.

Reading habits (and cigarette prices) have changed since 1935, but Penguin still believes in publishing the best books for everybody to enjoy. We still believe that good design costs no more than bad design, and we still believe that quality books published passionately and responsibly make the world a better place.

So wherever you see the little bird – whether it's on a piece of prize-winning literary fiction or a celebrity autobiography, political tour de force or historical masterpiece, a serial-killer thriller, reference book, world classic or a piece of pure escapism – you can bet that it represents the very best that the genre has to offer.

Whatever you like to read – trust Penguin.

read more
www.penguin.co.uk